Air Conditioning

JOHN DEERE

FUNDAMENTALS OF SERVICE
Air Conditioning
ISBN-0-86691-353-X

FOS5710NC (2012)   (ENGLISH)

Servicing, testing, and maintenance guide for air conditioning systems in off-road vehicles, trucks, buses, and automobiles

Deere & Company
LITHO IN U.S.A.

Introduction

Free Catalog — Call 1-800-522-7448

Check out all of our titles in the FUNDAMENTALS OF SERVICE series!

Here are a few of the titles in this series:

Hydraulics
Service, testing, and maintenance guide for hydraulic systems in off-road vehicles, trucks, and buses.

Shop Tools
A basic guide showing the right tool for each type of job and its proper use.

Electronic and Electrical Systems
Service, testing and maintenance guide for electronic and electrical systems in off-road vehicles, trucks, and buses.

Identification of Parts Failures
A highly illustrated failure analysis guide for automotive and off-road vehicle parts.

Engines
Service, testing, and maintenance guide for engine systems in off-road vehicles, trucks, and buses.

Welding
The fundamentals of welding, cutting, brazing, soldering, and surfacing of metals.

Power Trains
Service, testing, and maintenance guide for power trains in off-road vehicles, trucks, and buses.

Our FUNDAMENTALS OF SERVICE series brings together all the technical information you need and combines it with clearly written and amply illustrated instructional aids for many types of mechanical systems, their components, the tools needed, and repair procedures.

There are many ways to order, to inquire into prices, or to receive our free catalog:

- Call 1-800-522-7448 to order using a credit card
- Search online from http://www.JohnDeere.com/publications

Air Conditioning

Air Conditioning is the definitive "how-to" book of automotive air conditioning — from showing you how to diagnose problems and test components to explaining how to repair the system. And when we say "show you," we mean just that! Our book is filled with illustrations to clearly demonstrate what must be done ... photographs, drawings, pictorial diagrams, troubleshooting charts, tables, and diagnostic charts.

Instructions are written in simple language so that they can be easily understood. Starting with how air conditioning works, we build up to why it fails and what to do about it. This book can be used by anyone, from a novice to an experienced mechanic.

By starting with the basics, the book builds your knowledge step-by-step. Chapter 1 covers the basic principles of refrigeration. Chapter 2 covers the use of refrigerants, including how to properly evacuate, recycle, and dispose of them. In Chapter 3, the basic systems are introduced. The rest of the book covers testing, diagnosing, and servicing complete systems.

ACKNOWLEDGMENTS
John Deere gratefully acknowledges help from the following: American Society of Agricultural Engineers (ASAE); American Society of Heating, Refrigerating and Air-Conditioning Engineers (ASHRAE); Frigidaire, Division of General Motors Corp.; Harrison Radiator Division, General Motors Corp.; Hupp, Inc.; Industrial Educational Department, University of Texas; John E. Mitchell Company, Inc.; Kelvinator Division, American Motors Corp.; Nuday Co.; OTC Division, SPX Corporation; Red Dot Corporation; Society of Automotive Engineers (SAE); Stolper Industries; Technical Chemical Company; Tecumseh Products Co.; Warner Electric Clutch Co.; York, Division of Borg-Warner Corp.

Fundamentals of Service (FOS)

Fundamentals of Service (FOS) is a series of manuals created by Deere & Company. Each book in the series is conceived, researched, outlined, edited, and published by Deere & Company, John Deere Publishing. Authors are selected to provide a basic technical manuscript that could be edited and rewritten by staff editors.

HOW TO USE THE MANUAL: This FOS manual can be used by anyone — experienced mechanics, shop trainees, vocational students, and lay readers.

Persons not familiar with the topics discussed in this book should begin with Chapter 1 and then study the chapters in sequence. The experienced person can find what is needed on the "Contents" page.

Each guide was written by Deere & Company, John Deere Publishing staff in cooperation with the technical writers, illustrators, and editors at Almon, Inc. — a full-service technical publications company headquartered in Waukesha, Wisconsin (www.almoninc.com).

This material is the property of Deere & Company, John Deere Publishing. All use and/or reproduction not specifically authorized by Deere & Company, John Deere Publishing is prohibited.

Contents

Basics of Air Conditioning
Introduction..01-1
Basic Principles of Refrigeration...................01-2
States of Matter...01-2
Heat and Matter...01-3
Heat Movement..01-3
Pressure and Heat......................................01-4
Heat Measurement.....................................01-4
Latent Heat..01-5
Refrigerants...01-6
Basic Refrigeration Cycle 01-7
Heat and Pressure.....................................01-8
Removing Moisture....................................01-8
Summary: Basics of Air Conditioning............01-8
R-12 Temperature-Pressure Relation
   Chart — Low Side 01-9
R-134a Temperature-Pressure
   Relation Chart — Low Side.........................01-10
Test Yourself...01-11

Refrigerants and Oil
Introduction..02-1
Other Terms for Refrigerants 02-1
Pressure and Temperature Relationship 02-1
Handling Refrigerants.................................02-2
Safety Rules for Refrigerants 02-3
Moisture in the System................................02-4
Refrigeration Oil 02-5
Guidelines for Recycling Disposal....................02-5
Information Sources 02-6
Test Yourself...02-6

Basic System: How It Works
Introduction..03-1
Compressor..03-2
Compressor Relief Valve............................03-4
Superheat Shutoff Switch 03-5
High- and Low-Pressure Switches 03-6
Choice of Compressor................................03-6
Compressor Noise Complaints...................03-7
Condenser..03-7
Types of Condensers.................................03-8
Expansion Valve 03-9
Service Precautions...................................03-12
Evaporator..03-13
Fan Speed 03-14

Problems of Flooded or Starved
   Evaporator Coils....................................03-15
Receiver-Dryer (Dehydrator) 03-16
Accumulators..03-17
Use of Screens in the System 03-17
What Happens When Refrigerant Is Blocked ...03-17
Thermostat and Magnetic Clutch Systems.......03-18
Thermostat Control.....................................03-19
Magnetic Clutch..03-20
Bypass Systems..03-22
Suction Throttling Regulators 03-23
Lines and Connections...............................03-25
Test Yourself...03-26

Service Equipment
Introduction..04-1
Refrigerant Recovery and Recycling Station......04-2
Gauge and Manifold Set.............................04-3
Service Valves..04-6
Leak Detectors 04-9
Vacuum Pump..04-10
Other Service Tools 04-10
Test Yourself...04-10

Inspecting the System
Introduction..05-1
Visual Inspection of the System 05-2
Operating Inspection of the System 05-3
Test Yourself...05-4

Diagnosing the System
Introduction..06-1
Troubleshooting Customer Complaints 06-2
Flow Charts for Diagnosing the System 06-6
Condition 1...06-7
Condition 2...06-9
Condition 3...06-11
Condition 4...06-13
Diagnostic Chart...06-14
Test Yourself...06-14

Testing and Adjusting the System
Introduction..07-1
Installing Gauge Set to Check System
   Operation 07-1

Continued on next page

Original Instructions. All information, illustrations and specifications in this manual are based on the latest information available at the time of publication. The right is reserved to make changes at any time without notice.

COPYRIGHT © 2009
DEERE & COMPANY
Moline, Illinois
All rights reserved.
A John Deere ILLUSTRUCTION ® Manual
Previous Editions
Copyright © 1970, 1973, 1979, 1981, 1984, 1991, 1992, 1994, 2005

i

Contents

Page

Adding Refrigerant to the System.......................07-3
Volumetric Test of Compressor...........................07-4
Shaft Seal Leak Test...07-5
Checking and Adding Oil to
   Reciprocating Piston Compressors................07-6
Checking and Adding Oil to Axial
   Piston Compressors.......................................07-7
Bench Testing Expansion Valve for Efficiency....07-9
Thermostatic Temperature Control Switch 07-11
Adjusting Thermostat..07-11
Checking Clutch Coil for Electrical Operation...07-12
Leak Testing System Using Electronic
   Leak Detector..07-13
Test Yourself...07-14

Preparing System for Service
Introduction..08-1
Refrigerant Recovery..08-1
Evacuating System Using a Charging Station....08-2
Charging System Using Charging Station..........08-3
Charging System Using 15 Ounce (0.4
   L) Containers..08-4
Isolating Compressor from System 08-5
Test Yourself..08-6

Definitions and Conversions
Definitions of Terms and Symbols A-1
Measurement Conversion Chart.......................... A-4

Answers to Test Yourself Questions
Answers to Chapter 1 Questions......................... B-1
Answers to Chapter 2 Questions......................... B-1
Answers to Chapter 3 Questions......................... B-1
Answers to Chapter 4 Questions......................... B-1
Answers to Chapter 5 Questions......................... B-1
Answers to Chapter 6 Questions......................... B-1
Answers to Chapter 7 Questions......................... B-2
Answers to Chapter 8 Questions......................... B-2

Basics of Air Conditioning

Introduction

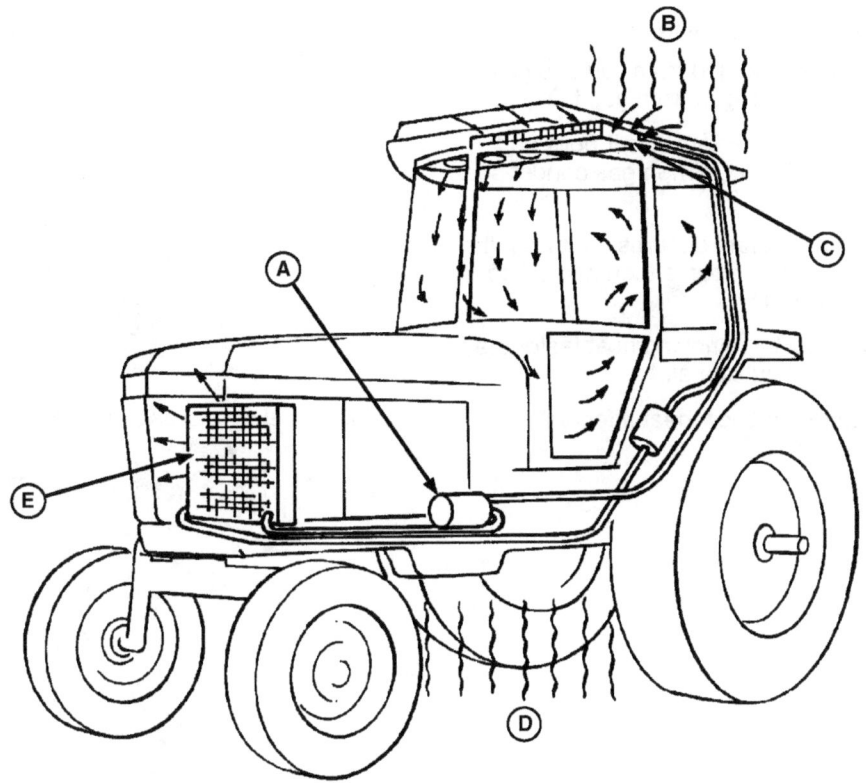

Fig 1. — What Air Conditioning Does

A—Compressor
B—Heat from Sun and Outside Air
C—Evaporator
D—Field or Road Heat
E—Condenser

Air conditioning (Fig. 1) removes heat from inside air and discharges the heat outside.

The temperature in the cab is reduced by removing heat faster than it comes in.

Let's look at the basic principles of air conditioning — laws of matter, heat, and refrigeration.

DK75838,0000093 -19-09OCT12-1/1

Basics of Air Conditioning

Basic Principles of Refrigeration

Air conditioners work on these principles:

- Liquids absorb heat when changed from liquid to gas.
- Gases give off heat when changed from gas to liquid.

The water **absorbs** heat from the flame as it boils and changes to a gas or vapor (Fig. 2). When gas condenses to a liquid, it **radiates** heat (Fig. 3).

In air conditioning, liquid refrigerant absorbs heat from the air when it changes to gas. This heat is then carried off and radiated into the outside air.

The temperature is kept cool by removing heat faster than it comes in from the sun and outside air.

Look at the states of matter and how heat affects them.

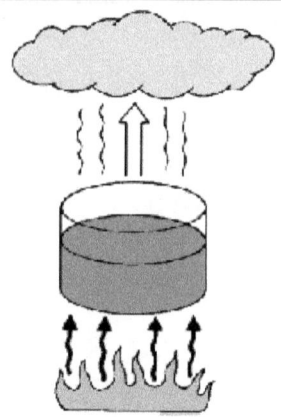

Fig. 2 — Basic Principles of Refrigeration — Liquid to Gas

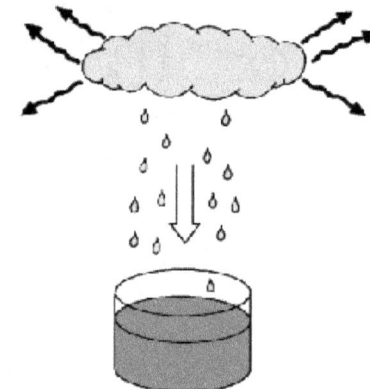

Fig. 3 — Basic Principles of Refrigeration — Gas to Liquid

States of Matter

  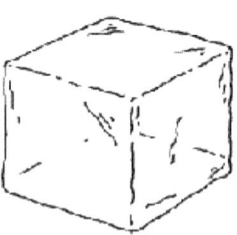

Fig. 4 — Three States of Water

All matter can be in any of three states (Fig. 4); gas, liquid, or solid. The steam rising from a heated kettle of water is familiar as a gas or vapor. Yet this vapor can be changed to a liquid by cooling. And, the liquid can be changed to a solid (ice) by further cooling.

Even hard steel tools can be changed to a liquid by heating them. Even more heat can change them from a liquid to a gas.

Continued on next page

Basics of Air Conditioning

All matter is composed of molecules that are moving in the mass, whatever its state (Fig. 5). The amount of molecule movement determines the density or solidity of matter. This is called the **theory of molecular motion.**

A—Vapor or Gas
B—Liquid
C—Solid

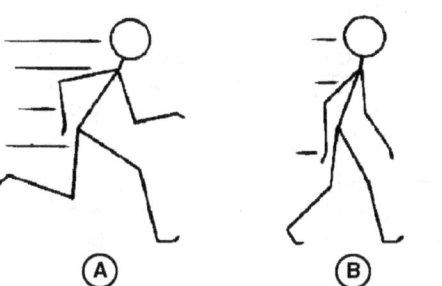

Fig. 5 — Movement of Molecules

Heat and Matter

Since all matter is composed of molecules in motion, **heat** becomes the controlling factor of molecular movement. The familiar state of all matter is the state we normally see it in, when no outside force is used to add or remove its heat.

Though we use the word "cold" constantly, cold is a relative term that refers only to **absence of heat.** Cold means that a substance contains less heat than another warmer substance. Cold as a complete absence of heat, in which all molecular action stops, has never been attained by man but is believed to be −459°F (−273°C). This is known as **absolute zero.**

Heat Movement

In an air conditioning system, heat must be removed from the passenger area to obtain a temperature that is comfortable to the occupants. This heat must then be expelled to the outside air. This is done by making use of the nature of heat movement.

Heat always moves away from the heat source (Fig. 6).

The rapidly moving molecules of the warmer substance impart some of their energy to the cooler, slower moving molecules. This slows down the molecules of the warmer substance and speeds up the molecules of the colder substance. This heat exchange can go on until the molecules of both substances are moving at the same rate; then their temperatures are the same and there is no further heat exchange between them.

In some heat exchanges, the molecules will change their **shape** instead of their speed of movement. This change of shape is caused by one or more atoms making up

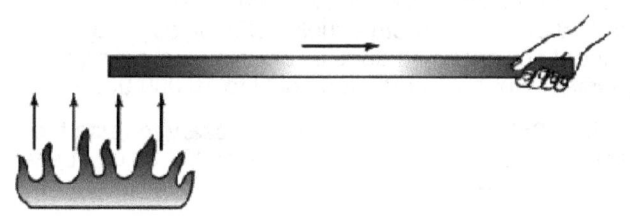

Fig. 6 — Heat Travels

the molecules changing position, which will cause the substance to change from a gas to a liquid, a liquid to a solid, and vice versa.

This molecular change is the basis for air conditioning with its exchange of heat energy between the gas and liquid states of its refrigerant liquid.

Pressure and Heat

The temperature at which a liquid boils will vary with the **pressure** on the liquid (Fig. 7). Decreasing the pressure **lowers** the boiling point, while increasing the pressure **raises** the boiling point.

For example: Water under 20 psi (139 kPa) (0.7 bar) pressure boils at 258°F (126°C), while water in a vacuum of 20 inches (508 mm) of mercury will boil at 160°F (71°C).

A—Open Atmospheric Pressure: 15 psi (103 kPa) (1.0 bar). Boils at 212°F (100°C).
B—Pressure: 20 psi (139 kPa) (1.4 bar). Boils at 258°F (126°C).
C—Partial Vacuum: 20 inches Hg (508 mm), 10 psi (69 kPa) (0.7 bar). Boils at 160°F (71°C).

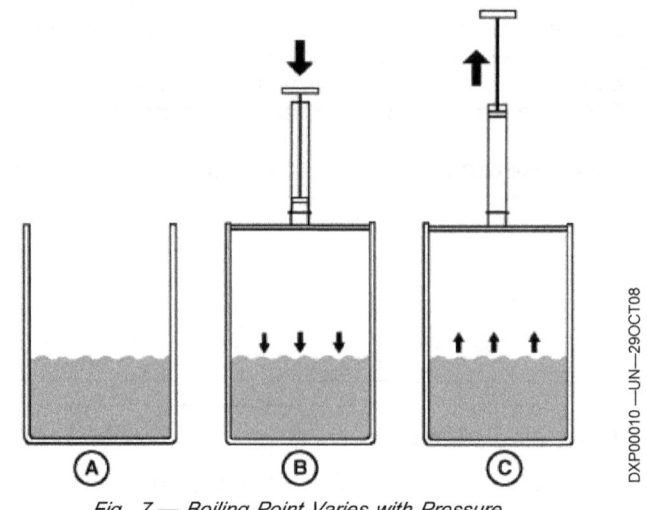

Fig. 7 — Boiling Point Varies with Pressure

Heat Measurement

Heat is measured in units of energy called British Thermal Units (Btu) (Fig. 8). One Btu is the amount of heat energy required to raise the temperature of one pound of water one degree Fahrenheit.

In the metric system of measurement, Joules are used instead of Btu. One Btu equals 1055 joules. A joule is the amount of work being done by one Newton, acting through a distance of one meter and equal to 10,000,000 ergs.

1 Btu (1055 J) equals the heat necessary to raise the temperature of 1 lb (0.45 kg) of water 1°F (1°C) at sea level pressure.

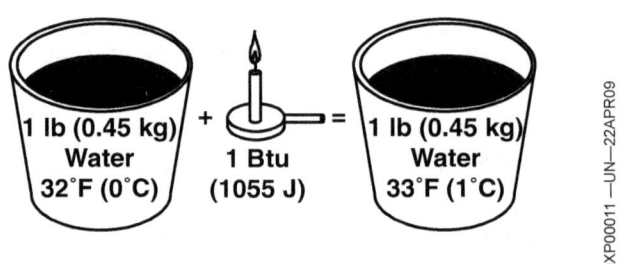

Fig. 8 — Heat Is Measured in Btu (Joules)

Latent Heat

Latent heat is the amount of energy necessary to change a substance from one state of matter to another without changing its temperature (Fig. 9).

We know that ice will change to water at 32°F (0°C). Each pound of ice requires 143 Btu to make this change. (Each gram of ice requires 333 J.)

To change water into steam at 212°F (100°C), each pound of water must have 970 Btu added to it. (Each gram of water must have 2260 J added.)

We call the heat that must be added to ice to cause a change of state the **latent heat of liquidization.** We call the heat that must be added to water to cause it to change its state the **latent heat of vaporization.**

If we reverse this change of state, steam, with a heat intensity of 212°F (100°C), will give off 970 Btu per pound (2260 J per gram) as it condenses to water. This heat release is called the **latent heat of condensation.**

As the water is further cooled, the molecules are realigned and change the liquid to a solid (ice). The heat given off then is called the **latent heat of freezing**, which will be 143 Btu of heat per pound (333 kJ per kg) at 32°F (0°C).

This principle is the basis for air conditioning operation. A refrigerant is chosen for its ability to change its state readily and give off or absorb Btu (joules).

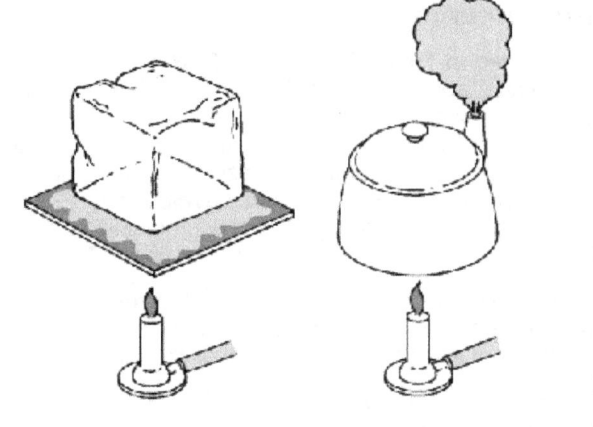

Fig. 9 — Latent Heat

Following is a list giving the latent heat of vaporization of some of the common refrigerants as compared to water:

Substance	Btu per Pound (Joules per Gram)
Water	970 at 212°F (2260 at 100°C)
Ammonia	565 at 5°F (1310 at −15°C)
Sulphur Dioxide	169 at 5°F (390 at −15°C)
Refrigerant-12	69 at 5°F (160 at −15°C)

Refrigerants

Introduced in 1930, Refrigerant-12 (R-12) proved to be the most stable and easiest to handle. However, due to its chlorofluorocarbon (CFC) base, research has shown that it has a harmful environmental effect by contributing to the depletion of the atmosphere's ozone layer. A more recent Refrigerant-134a (R-134a) was developed to replace R-12. This R-134a refrigerant is ecologically safer for the environment and has phased out the use of R-12.

It must be noted that R-12 and R-134a refrigerants are **not** compatible. They cannot be used together or in a system not specifically designed for their intended use.

A low boiling point of –22°F (–30°C) for R-12 and –15°F (–26°C) for R-134a demonstrates their capability to remove large quantities of heat from the surrounding air.

Fig. 10 shows how a refrigerant removes heat from the air. A container of liquid refrigerant placed in an insulated box will boil furiously at room temperature and atmospheric pressure. As this change of state occurs, heat will follow its natural tendency and move from the warmer air inside the box to the boiling refrigerant. This heat will then be carried out of the box with the refrigerant vapors, lowering the temperature of the air inside the box. This could continue until the air and the refrigerant are both at –22°F (–30°C) for R-12 or –15°F (–26°C) for R-134a, when the refrigerant would stop boiling. At this time the heat content of both substances would be equal and there would be no further heat transfer between the air and the refrigerant.

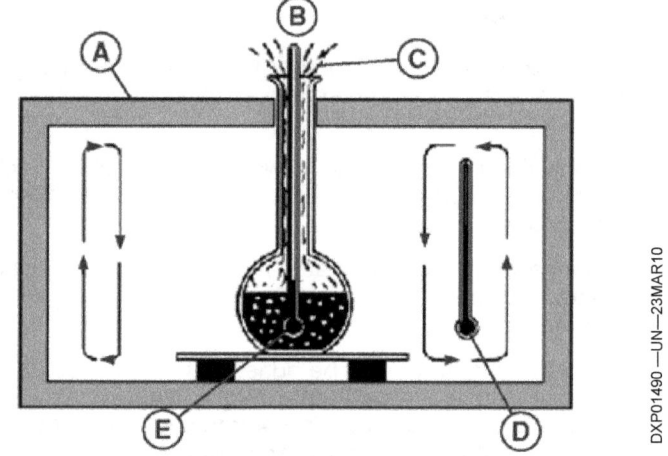

Fig. 10 — How Refrigerant Ion Is Produced by Evaporating Refrigerant

A—Insulated Box
B—Atmosphere at 75°F (24°C)
C—Refrigerant Vapor
D—Air at 30°F (–1°C)
E—Boiling R-12 at –22°F (–30°C)

We will look at refrigerants more closely in "Refrigerants and Oil" on page 02-1.

Basic Refrigeration Cycle

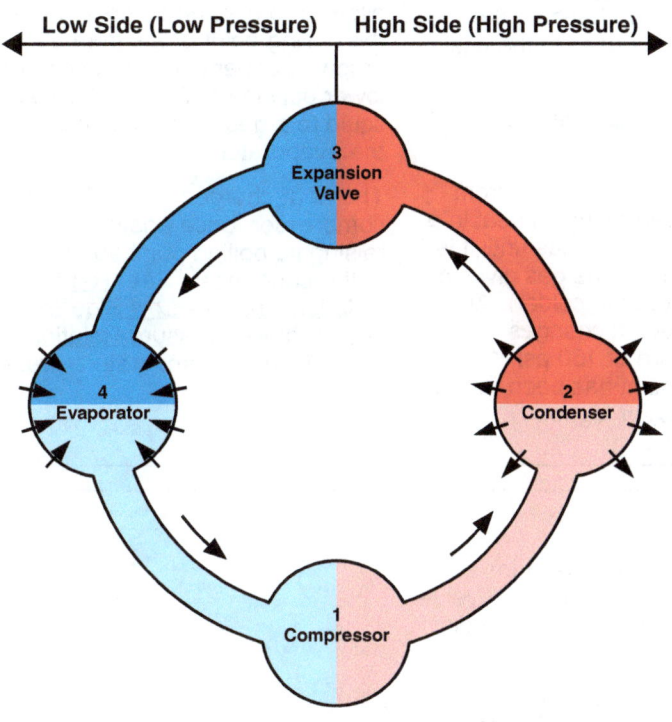

Fig. 11 — Complete Four-Part Cycle of Refrigeration

A—High-Pressure Liquid
B—High-Pressure Gas
C—Low-Pressure Liquid
D—Low-Pressure Gas

For an air conditioning system to operate economically, the refrigerant must be used repeatedly. For this reason, all air conditioners use the same cycle of **compression, condensation, expansion,** and **evaporation** in a closed circuit (Fig. 11). The same refrigerant is used to move the heat from one area, to cool this area, and to expel this heat in another area.

In Fig. 11, note that the four-part cycle is divided at the center into a **high side** and a **low side.** This refers to the pressures of the refrigerant in each side of the system.

To understand the basic refrigerant cycle, let's start at the **compressor** (1). The refrigerant comes into the compressor as a low-pressure gas, is compressed, and moves out of the compressor as a high-pressure gas.

The gas then flows to the **condenser** (2). Here the gas condenses to a liquid, giving off its heat to the outside air.

The liquid then moves to the **expansion valve** (3) under high pressure. This valve restricts the flow of the fluid, thus lowering its pressure as it leaves the expansion valve.

The low-pressure liquid then moves to the **evaporator** (4), where heat from the inside air is absorbed and changes it from a liquid to a gas.

As a hot low-pressure gas, the refrigerant moves to the compressor (1), where the entire cycle is repeated.

Heat and Pressure

As we explained earlier, the boiling point of a liquid (changing it to a gas) can be varied by changing the pressure on the liquid. This principle is used in the basic refrigeration cycle.

Liquids will not compress; however, a gas can be compressed.

In the refrigeration system (Fig. 11), the cool gas from the evaporator is compressed, concentrating its heat in a small area. This compression heats up the gas until it is warmer than the outside air. As a result, the gas gives off its heat to the outside air (moving hotter to colder). As the gas gives off its heat at the condenser, it changes from a gas to a liquid. Under a high pressure of 160 psi (1103 kPa) (11.0 bar) and up, the boiling point has been raised from −22°F to 120°F (−30°C to 49°C) or higher.

As the liquid reaches the expansion valve and is metered to the low side of the system, the liquid refrigerant is allowed to expand as pressure is removed from it. Here the boiling point drops from 80°F (27°C) to 30°F (−1°C) or lower, depending on the controls of the system. At this lower boiling point, the refrigerant changes again from a liquid to a gas, absorbing heat from the inside air through the evaporator.

The heat-charged gas is then compressed by the compressor, once again, concentrating its heat and raising its boiling point so that heat can be exchanged in the condenser. See "R-12 Temperature-Pressure Relation Chart — Low Side" on page 01-9 or "R-134a Temperature-Pressure Relation Chart — Low Side" on page 01-10 for temperature-pressure relation charts.

Removing Moisture

The principle of heat and pressure is also used in removing harmful moisture from the system, but in reverse order — we can just as effectively lower the boiling point by decreasing the pressure.

As the system is pumped down to a vacuum (Fig. 12), moisture will vaporize and be carried out by the vacuum pump. In other words, the water boils at a lower temperature because of the reduced pressure and so vaporizes and is drawn off.

A—Water at Normal Atmospheric Pressure (15 psi [103 kPa] [1.0 bar]) Boils at 212°F (100°C)

B—Water at Partial Vacuum (20 Inches Hg [508 mm], 10 psi [69 kPa] [0.7 bar]) Boils at 160°F (71°C)

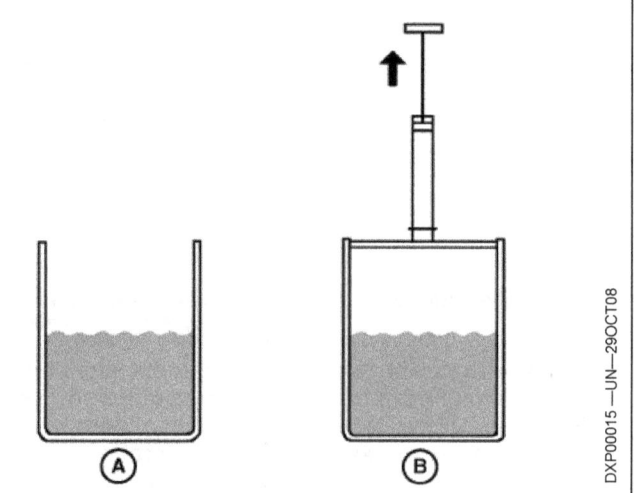

Fig. 12 — Boiling of Liquid Under a Decrease in Pressure

Summary: Basics of Air Conditioning

Here is a review of the basic principles:

- Liquids absorb heat when changed from liquid to gas.
- Gases give off heat when changed from gas to liquid.
- Heat always moves from the hotter to the colder.
- Temperature at which a liquid changes to a gas varies with the pressure on it.
- Refrigerant liquids must have a low boiling point and heat and cool readily for best heat exchanges.
- Basic refrigeration cycle is compression, condensation, expansion, and evaporation:
  a. Compression heats up the gas.
  b. Condensation changes gas to liquid and releases heat.
  c. Expansion reduces pressure.
  d. Evaporation changes liquid to gas and absorbs heat.

R-12 Temperature-Pressure Relation Chart — Low Side

Refrigerant-12											
Temp.		Pressure			Temp.		Pressure			Temp.		Pressure		
°F.	°C.	PSI	kPa	bar	°F.	°C.	PSI	kPa	bar	°F.	°C.	PSI	kPa	bar
0	-18	9.1	62.7	0.63	50	10	46.6	321.3	3.21	90	32	99.6	686.7	6.87
2	-17	10.1	69.6	0.70	51	11	47.8	329.6	3.30	91	33	101.3	698.5	6.98
4	-16	11.2	77.2	0.77	52	11	48.7	335.8	3.36	92	33	103.0	710.2	7.10
6	-14	12.3	84.8	0.85	53	12	49.8	343.4	3.43	93	34	104.6	721.2	7.21
8	-13	13.4	92.4	0.92	54	12	50.9	351.0	3.51	94	34	106.3	732.9	7.33
10	-12	14.6	100.7	1.01	55	13	52.0	358.5	3.59	95	35	108.1	745.3	7.45
12	-11	15.8	108.9	1.09	56	13	53.1	366.1	3.66	96	36	109.8	757.1	7.57
14	-10	17.1	117.9	1.18	57	14	55.4	382.0	3.82	97	36	111.5	768.8	7.69
16	-9	18.3	126.2	1.26	58	14	56.6	390.3	3.90	98	37	113.3	781.2	7.81
18	-8	19.7	135.8	1.36	59	15	57.1	393.7	3.94	99	37	115.1	793.6	7.94
20	-7	21.0	144.8	1.45	60	16	57.7	397.8	3.98	100	38	116.9	806.0	8.06
21	-6	21.7	149.6	1.50	61	16	58.9	406.1	4.06	101	38	118.8	819.1	8.19
22	-6	22.4	154.4	1.54	62	17	60.0	413.7	4.14	102	39	120.6	831.5	8.32
23	-5	23.1	159.3	1.59	63	17	61.3	422.7	4.23	103	39	122.4	843.9	8.44
24	-4	23.8	164.1	1.64	64	18	62.5	430.9	4.31	104	40	124.3	857.0	8.57
25	-4	24.6	169.6	1.70	65	18	63.7	439.2	4.39	105	41	126.2	870.1	8.70
26	-3	25.3	174.4	1.74	66	19	64.9	447.5	4.47	106	41	128.1	883.2	8.83
27	-3	26.1	180.0	1.80	67	19	66.2	456.4	4.56	107	42	130.0	896.4	8.96
28	-2	26.8	184.8	1.85	68	20	67.5	465.4	4.65	108	42	132.1	910.8	9.11
29	-2	27.6	190.3	1.90	69	21	68.8	474.4	4.74	109	43	135.1	931.5	9.32
30	-1	28.4	195.8	1.96	70	21	70.1	483.3	4.83	110	43	136.0	937.7	9.38
31	-1	29.2	201.3	2.01	71	22	71.4	492.3	4.92	111	44	138.0	951.5	9.52
32	0	30.0	206.9	2.07	72	22	72.8	502.0	5.02	112	44	140.1	966.0	9.66
33	1	30.9	213.1	2.13	73	23	74.2	511.6	5.12	113	45	142.1	979.8	9.80
34	1	31.7	218.6	2.19	74	23	75.5	520.6	5.21	114	46	144.2	994.3	9.94
35	2	32.5	224.1	2.24	75	24	76.9	530.2	5.30	115	46	146.3	1008.7	10.09
36	2	33.4	230.3	2.30	76	24	78.3	539.9	5.40	116	47	148.4	1023.2	10.23
37	3	34.3	236.5	2.36	77	25	79.2	546.1	5.46	117	47	151.2	1042.5	10.43
38	3	35.1	242.0	2.42	78	26	81.1	559.2	5.59	118	48	152.7	1052.9	10.52
39	4	36.0	248.2	2.48	79	26	82.5	568.8	5.69	119	48	154.9	1068.0	10.68
40	5	36.9	254.4	2.54	80	27	84.0	579.2	5.79	120	49	157.1	1083.2	10.83
41	5	37.9	261.3	2.61	81	27	85.5	589.5	5.90	121	49	159.3	1093.4	10.98
42	6	38.8	267.5	2.68	82	28	87.0	599.9	6.00	122	50	161.5	1113.5	11.14
43	6	39.7	273.7	2.74	83	28	88.5	610.2	6.10	123	51	163.8	1129.4	11.29
44	7	40.7	280.6	2.81	84	29	90.1	621.2	6.21	124	51	166.1	1145.3	11.45
45	7	41.7	287.5	2.88	85	29	91.7	632.3	6.32	125	52	168.4	1161.1	11.61
46	8	42.6	293.7	2.94	86	30	93.2	642.6	6.43	126	52	170.7	1177.0	11.77
47	8	43.6	300.6	3.01	87	31	94.8	653.6	6.54	127	53	173.1	1193.5	11.94
48	9	44.6	307.5	3.08	88	31	96.4	664.7	6.65	128	53	175.4	1209.4	12.09
49	9	45.6	314.4	3.14	89	32	98.0	675.7	6.76	129	54	177.8	1225.9	12.26
										130	54	182.2	1256.3	12.56
										131	55	182.6	1259.0	12.59
										132	56	185.1	1276.3	12.76
										133	56	187.6	1293.5	12.94
										134	57	190.1	1310.7	13.11

Fig. 13 — Temperature-Pressure Relation Chart (R-12)

R-134a Temperature-Pressure Relation Chart — Low Side

Refrigerant-134a											
Temp.		Pressure			Temp.		Pressure			Temp.		Pressure		
°F.	°C.	PSI	kPa	bar	°F.	°C.	PSI	kPa	bar	°F.	°C.	PSI	kPa	bar
0	-18	6.2	43.1	0.43	46	8	40.6	279.7	2.80	92	33	108.0	744.8	7.45
1	-17	6.7	46.5	0.46	47	8	41.6	287.0	2.87	93	34	110.0	758.2	7.58
2	-17	7.2	49.9	0.50	48	9	42.7	294.4	2.94	94	34	111.9	771.8	7.72
3	-16	7.8	53.5	0.54	49	9	43.8	301.9	3.02	95	35	113.9	785.6	7.85
4	-16	8.3	57.1	0.57	50	10	44.9	309.6	3.10	96	36	116.0	799.5	8.00
5	-15	8.8	60.7	0.61	51	11	46.0	317.3	3.17	97	36	118.0	813.6	8.14
6	-14	9.3	64.5	0.64	52	11	47.2	325.2	3.25	98	37	120.1	827.8	8.28
7	-14	9.9	68.3	0.68	53	12	48.3	333.1	3.33	99	37	122.2	842.3	8.43
8	-13	10.5	72.1	0.72	54	12	49.5	341.2	3.41	100	38	124.3	856.8	8.57
9	-13	11.0	76.1	0.76	55	13	50.7	349.4	3.50	101	38	126.4	871.6	8.72
10	-12	11.6	80.1	0.80	56	13	51.9	357.8	3.58	102	39	128.6	886.5	8.87
11	-12	12.2	84.2	0.84	57	14	53.1	366.2	3.66	103	39	130.8	901.6	9.02
12	-11	12.8	88.3	0.88	58	14	54.4	374.8	3.75	104	40	133.0	916.8	9.17
13	-11	13.4	92.5	0.92	59	15	55.6	383.5	3.83	105	41	135.2	932.3	9.32
14	-10	14.0	96.8	0.97	60	16	56.9	392.3	3.92	106	41	137.5	947.9	9.48
15	-9	14.7	101.2	1.01	61	16	58.2	401.2	4.01	107	42	139.8	963.7	9.64
16	-9	15.3	105.7	1.05	62	17	59.5	410.2	4.10	108	42	142.1	979.6	9.80
17	-8	16.0	110.2	1.10	63	17	60.8	419.4	4.19	109	43	144.4	995.7	9.96
18	-8	16.7	114.8	1.15	64	18	62.2	428.7	4.29	110	43	146.8	1012.1	10.12
19	-7	17.3	119.5	1.19	65	18	63.5	438.1	4.38	111	44	149.2	1028.5	10.29
20	-7	18.0	124.3	1.24	66	19	64.9	447.7	4.47	112	44	151.6	1045.2	10.45
21	-6	18.7	129.1	1.29	67	19	66.3	457.4	4.57	113	45	154.0	1062.1	10.62
22	-6	19.4	134.1	1.34	68	20	67.8	467.2	4.67	114	46	156.5	1079.1	10.79
23	-5	20.2	139.1	1.39	69	21	69.2	477.2	4.77	115	46	159.0	1096.3	10.96
24	-4	20.9	144.2	1.44	70	21	70.7	487.2	4.87	116	47	161.5	1113.7	11.14
25	-4	21.7	149.4	1.50	71	22	72.1	497.4	4.97	117	47	164.1	1131.3	11.31
26	-3	22.4	154.6	1.54	72	22	73.6	507.8	5.07	118	48	166.7	1149.1	11.49
27	-3	23.2	160.0	1.60	73	23	75.2	518.3	5.18	119	48	169.3	1167.0	11.67
28	-2	24.0	165.4	1.65	74	23	76.7	528.9	5.29	120	49	171.9	1185.2	11.85
29	-2	24.8	171.0	1.71	75	24	78.3	539.7	5.40	121	49	174.5	1203.5	12.03
30	-1	25.6	176.6	1.77	76	24	79.8	550.6	5.50	122	50	177.2	1222.0	12.22
31	-1	26.4	182.3	1.82	77	25	81.4	561.6	5.61	123	51	180.0	1240.8	12.41
32	0	27.3	188.1	1.88	78	26	83.1	572.8	5.73	124	51	182.7	1259.7	12.60
33	1	28.1	194.0	1.94	79	26	84.7	584.1	5.84	125	52	185.5	1278.8	12.79
34	1	29.0	200.0	2.00	80	27	86.4	595.6	5.96	126	52	188.3	1298.1	12.98
35	2	29.9	206.1	2.06	81	27	88.1	607.2	6.07	127	53	191.1	1317.6	13.18
36	2	30.8	212.3	2.12	82	28	89.8	618.9	6.19	128	53	194.0	1337.3	13.38
37	3	31.7	218.6	2.19	83	28	91.5	630.8	6.31	129	54	196.8	1357.2	13.57
38	3	32.6	225.0	2.25	84	29	93.2	642.9	6.43	130	54	199.8	1377.3	13.78
39	4	33.6	231.4	2.32	85	29	95.0	655.1	6.55	131	55	202.7	1397.6	13.98
40	4	34.5	238.0	2.38	86	30	96.8	667.4	6.67	132	56	205.7	1418.1	14.18
41	5	35.5	244.7	2.45	87	31	98.6	680.0	6.80	133	56	208.7	1438.8	14.39
42	6	36.5	251.5	2.52	88	31	100.5	692.6	6.93	134	57	211.7	1459.7	14.60
43	6	37.5	258.4	2.59	89	32	102.3	705.4	7.05					
44	7	38.5	265.4	2.65	90	32	104.2	718.4	7.18					
45	7	39.5	272.5	2.72	91	33	106.1	731.5	7.32					

Fig. 14 — Temperature-Pressure Relation Chart (R-134a)

Test Yourself

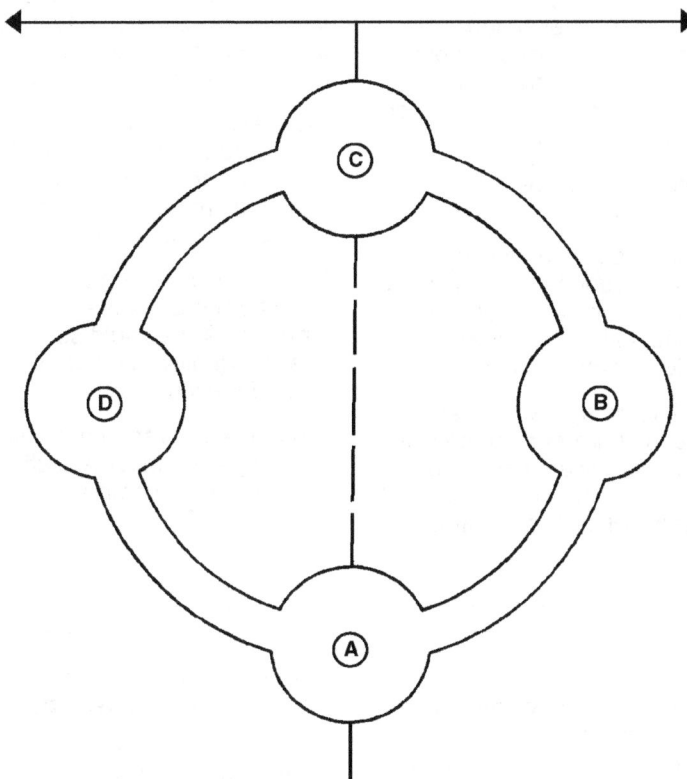

Fig. 15 — Diagram of Basic Refrigeration Cycle (See Question 6)

Questions

1. When a liquid is changed to a gas, does it absorb or give off heat?

2. If you increase the pressure on a liquid, does it raise or lower the boiling point?

3. (Fill in the blanks with "colder" or "hotter.") Heat always moves from the _____ to the _____.

4. Define a British Thermal Unit (Btu).

5. What is the biggest factor for using Refrigerant-134a over Refrigerant-12?

6. Mark up Fig. 15, "Diagram of Basic Refrigeration Cycle" to show the basic refrigeration cycle:

a. Label parts A, B, C, and D.

b. Show "high side" and "low side" by each arrow at the top.

c. Between each part of the cycle, label the refrigerant flow as high or low pressure, and as gas or liquid.

d. Show by arrows around parts "B" and "D" whether heat is absorbed or given off.

See "Answers to Chapter 1 Questions" on page B-1.

Refrigerants and Oil

Introduction

While **Refrigerant-12** was the most widely used refrigerant for many years, it has been phased out of use by **Refrigerant-134a**. This is due in large part to R-12's adverse effect on the earth's environment. The depletion of the earth's protective ozone layer mandates this change.

Ozone in the upper atmosphere protects the earth's inhabitants from the sun's ultraviolet rays.

The earth's inhabitants are subjected to increasing amounts of lethal radiation as the ozone layer in the upper atmosphere diminishes. This radiation can result in skin cancer, immune system damage, cataracts, as well as damage to food crops, plant life, and animal life.

It is imperative to recognize which refrigerant is used when servicing a system since they are **not** compatible. If there is even a trace of R-12 in a system designed for R-134a, a sludge forms, preventing lubricant flow through the system, resulting in compressor failure and damage to other system components.

Recovery and recycling systems **are required** for safe handling of these compounds and to prevent their exposures to the atmosphere. Each refrigerant requires its own recovery and recycling station.

There are federal EPA (Environmental Protection Agency) laws, regulations, and guidelines to ensure that proper recovery and recycling methods are followed. For additional information, see "Guidelines for Recycling Disposal" on page 02-5.

R-134a has a boiling point of –15°F (–26°C) while Rt-12 has a boiling point of –22°F (–30°C), which means that a container of the liquid sitting on a block of ice will boil. This is a quality needed for a refrigerant, which must work as a heat exchanger.

Another refrigerant, called **Refrigerant-22** (R-22), is generally used in large stationary systems equipped with a reciprocating compressor. R-22 has a boiling point of –46°F (–43°C).

Other Terms for Refrigerants

"Freon" has become a common shop term when referring to Refrigerant-12. However, "Freon" and "Freon-12" are registered trademarks of E. I. du Pont de Nemours and Company and should be used only when referring to refrigerants packaged and sold by Du Pont.

Refrigerant-12 is also packaged in the United States under several other brand names such as Genatron-12, Isotron-12, Ucon-12, and others.

Refrigerant-12 has been abbreviated as **R-12** and is accepted in the industry by this name.

KLEA-134A is a trademark of the ICI Corporation and SUVA-134A is a trademark of E. I. du Pont de Nemours and Company.

Refrigerant-134a is accepted in the industry by the name of **R-134a**.

Pressure and Temperature Relationship

In air conditioning a compartment, the objective is to allow the evaporator to reach its coldest point without icing up. Since ice will form at 32°F (0°C), the fins and cooling coils of the evaporator must not be allowed to drop below this point. Because of the temperature rise through the walls of the cooling fins and coils, the temperature of the refrigerant may be several degrees cooler than that of the air passing through the evaporator.

A thermometer should be used with a gauge and manifold set to properly service an air conditioner. See "Service Equipment" on page 04-1.

Handling Refrigerants

⚠ **CAUTION: Refrigerants must be handled with care to avoid danger.**

Do not discharge refrigerant into the atmosphere. By law, chlorofluorocarbon (CFC) refrigerants can no longer be released into the environment. They must be recovered by licensed air conditioner services. They will either return refrigerant to your air conditioner during repair, or send it to a recycling operation. The most recent air conditioning systems use a hydro fluorocarbon refrigerant (HFC R-134a) that will not harm the ozone layer of the environment.

⚠ **CAUTION: Liquid refrigerant, if allowed to strike the eye, can cause blindness. If allowed to strike the body, it can cause frostbite.**

If a refrigerant container is heated, or contacts a heating element, the pressure inside can build up and cause the container to explode.

If refrigerant is allowed to contact an open flame or heated metal, a poisonous gas will be created. Inhaling this gas can cause violent illness.

Refrigerant can be dangerous. It should only be handled by a certified service technician.

Safety Rules for Refrigerants

Fig. 1 — Safety Symbols for Refrigerants

A—Wear Eye Protection
B—Avoid Breathing Refrigerant
C—Potential Thermal Explosion
D—Keep Away from Open Flame
E—Wear Gloves
F—Recover Refrigerant (Refrigerant is Pressurized in System)

Safety symbols used for refrigerant are shown in (Fig. 1). Observe the following safety rules when handling refrigerant and refrigerant containers.

1. Do not handle refrigerant without suitable eye protection. Escaping refrigerant that might come in contact with the eyes can result in serious frostbite or blindness because of its low boiling point. The eyes should be washed immediately with clean, cold water for at least ten minutes if an accident occurs. Then go to a doctor or an eye specialist as soon as possible. The fluid you see escaping from open connections is oil, not refrigerant. But, it can also be harmful to the eyes because its dryness can dehydrate the tender tissues of the eyeball. Refrigerant can also escape, depending on where the open connection is.

2. Avoid breathing refrigerant and lubricant vapor or mist. Exposure may irritate nose and throat. Use service equipment certified to meet requirements of SAE J1990 (R-12) and SAE J2210 (R-134a) to remove and recover refrigerant from the system. If accidental system discharge occurs, ventilate area before resuming service. Additional health and safety information may be obtained from refrigerant and lubricant manufacturers.

3. R-134a service equipment or vehicle systems should not be pressure- or leak-tested with compressed air. Some mixtures of air and R-134a have been combustible at elevated pressures. These mixtures are potentially dangerous and may result in fire or explosion causing injury or property damage.

Additional health and safety information may be obtained from refrigerant and lubricant manufacturers.

4. Do not overheat the refrigerant container. Refrigerants and their containers heat quickly, causing a rapid and dangerous pressure buildup. This can be extremely dangerous in one- and two-pound containers having very thin walls.

Also, during the charging process, water temperature for heating the refrigerant containers should not exceed approximately 125°F (52°C). Higher temperatures will cause excessive pressure in the container which could explode.

5. Recover — do **NOT** discharge — refrigerant. Refrigerant subjected to open flame will result in phosgene gas, which is deadly.

6. Do not add anything but pure refrigerant and refrigerant oil into the system. Any additional compound can contain foreign substances not compatible with the chemical makeup of refrigerant, causing it to become chemically unstable and to lose its good qualities.

7. Do not handle damp containers with bare hands while charging a system. Frost will form on the outside of the container when it is wet and cause the hand to be frozen to the container. This can occur when using warm water to heat the container. If this happens, wet the container, and your hand will thaw and be released.

8. Do not weld or steam clean near or on an air conditioning system. Excessive pressure could build up in the system.

Continued on next page

Refrigerants and Oil

9. Before loosening a refrigerant fitting, refrigerant must be recovered.

10. When charging a system with the engine running, be sure the high-pressure gauge valve is **closed**.

Moisture in the System

Any air conditioning system should be as dry as it is possible to maintain it. The refrigerant has had virtually all the moisture removed from it during manufacture. Any moisture introduced into the system must come from outside sources such as a break in a line or from improper sealing of connections when a unit is removed for service.

Refrigerant will absorb moisture readily when exposed to it. To keep the system as moisture-free as possible, all systems use a receiver-dryer or accumulator containing a desiccant which can absorb great quantities of moisture. However, the desiccant can only absorb a predetermined amount and when the saturation point has been reached, the effectiveness is lost.

Moisture in greater concentrations than 14 parts per million causes damage to the inside parts of the system. Water reacts with R-12 to form hydrochloric acid and hydrofluoric acid. The greater the moisture content in the system, the more concentrated this corrosive acid. The acid eats away on all internal metal parts — iron, copper, and aluminum — releasing oxides into the refrigerant as foreign substances that seriously affect the ability of the refrigerant to absorb and give off heat.

Moisture is the greatest enemy of the air conditioning system. Once a system is saturated for a long time, irreparable damage is done inside the system. Pinholes can develop in the evaporator and condenser coils, ultimately requiring their replacement. Aluminum parts will be eaten away, ruining the compressor, while valves and fittings can be so corroded that they are no longer usable.

The complete removal of moisture from a system can be a serious problem for the service technician. Vacuum is used as the best means of removing moisture. Moisture forms in small droplets throughout the system at zero pressure or higher. Remembering that at zero pressure the system still has at least partial atmospheric pressure, as we pump the system down into a vacuum, pressure is released from the moisture droplets, changing them from a liquid to a vapor. The vacuum pump then removes the vaporized moisture. The inches (kPa) (bar) of vacuum reached plus the length of time the system is subjected to a vacuum will determine the amount of moisture removed.

The use of a vacuum pump (Fig. 2) is necessary for moisture removal.

Fig. 2 — Using a Vacuum Pump to Remove Moisture from the System

Do not use the compressor as a vacuum pump. The compressor is cooled and lubricated by refrigerant and oil, but when vacuuming, neither of these is done. The risks are high, so avoid the practice.

Any system seriously contaminated with moisture must have the receiver-dryer or accumulator replaced and pull a deep vacuum of 29 in. (737 mm) Hg minimum for 30 minutes.

Systems that are seriously contaminated with moisture should have the receiver-dryer or accumulator replaced prior to pumping down. A good vacuum pump will reduce moisture to a small percentage. The remaining moisture may then be readily absorbed by the desiccant. Complete dehydration of the system will then allow the unit to cool ten or more degrees.

Refrigeration Oil

Refrigerant oil is a special oil used in refrigeration systems. It is needed to lubricate the seals, gaskets, and other moving parts of the compressor. A small amount of oil is circulated through the system with the refrigerant. This also aids in keeping the thermostatic expansion valve in proper operating condition.

Only non-foaming oil specifically formulated for use in each air conditioner should be used.

Refrigeration oil used with R-12 is highly refined. It is a mineral oil with all impurities such as wax, moisture, and sulfur removed. **Never use a motor oil, regardless of grade, in an air conditioning system.**

Refrigeration oil is available in several grades or viscosities. Viscosity is determined by the time in seconds it takes a definite quantity of oil to flow through a certain size orifice at 100°F (38°C).

The lubricant used with R-134a is called polyalkalene glycol (PAG). It is a synthetic oil and is not compatible with R-12 refrigeration oil.

IMPORTANT: Avoid damaging the system — do not mix refrigerants. If R-12 is used in a R-134a system, the R-12 will not carry the lubricant through the system to other system components.

Check the oil level of the compressor each time the air conditioner is serviced. Always check the compressor manufacturer's recommendations before adding oil to the system. See "Checking and Adding Oil to Reciprocating Piston Compressors" on page 07-6.

Except when pouring, never allow the oil container to remain uncapped. Always be sure that the cap is in place and is tight. Oil absorbs moisture, and moisture is damaging to the system.

Handling Refrigeration Oil

Here are a few simple rules to follow when handling refrigeration oil:

- Use only approved refrigeration oil.
- Do not transfer oil from one container to another.
- Make sure the cap is tight on the container when not in use.
- Replace old oil if there is any doubt about its condition.
- Avoid contaminating the oil.

Guidelines for Recycling Disposal

By properly recycling or disposing of waste, you can help protect the health of your family, friends, and fellow workers, and future generations.

Be sure to check and follow federal, state, and local requirements (Fig. 3). Regulations, facilities, and guidelines for waste recycling and disposal are being revised, expanded, and publicized more each year.

You must learn about regulations, facilities, and guidelines pertaining to safe handling of refrigerants as well as safe waste disposal for all materials.

Fig. 3 — Check Your Telephone Directory for Organizations to Advise on Waste Disposal

Information Sources

There are federal EPA (Environmental Protection Agency) laws, regulations, and guidelines for waste recycling and disposal. However, state and county governments also have special regulations and programs. Be sure to know what they are. For more information, call the EPA Hotline: 1-800-424-9346.

EPA Regional Offices

Region 1: Connecticut, Maine, Massachusetts, New Hampshire, Rhode Island, Vermont

Environmental Protection Agency
One Congress Street, Suite 1100
Boston, MA 02114-2023
http://www.epa.gov/region01/
Telephone: (617) 918-1111

Region 2: New York, New Jersey, Puerto Rico, Virgin Islands

Environmental Protection Agency
290 Broadway
New York, NY 10007-1866
http://www.epa.gov/region02/
Telephone: (212) 637-3000

Region 3: Delaware, District of Columbia, Maryland, Pennsylvania, Virginia, West Virginia

Environmental Protection Agency
1650 Arch Street
Philadelphia, PA 19103-2029
http://www.epa.gov/region03/
Telephone: (215) 814-5000

Region 4: Alabama, Florida, Georgia, Kentucky, Mississippi, North Carolina, South Carolina, Tennessee

Environmental Protection Agency
Atlanta Federal Center
61 Forsyth Street, SW
Atlanta, GA 30303-3104
http://www.epa.gov/region04/
Telephone: (404) 562-9900

Region 5: Illinois, Indiana, Michigan, Minnesota, Ohio, Wisconsin

Environmental Protection Agency
77 W. Jackson Boulevard
Chicago, IL 60604-3507
http://www.epa.gov/region05/
Telephone: (312) 353-2000

Region 6: Arkansas, Louisiana, New Mexico, Oklahoma, Texas

Environmental Protection Agency
Fountain Place 12th Floor, Suite 1200
1445 Ross Avenue
Dallas, TX 75202-2733
http://www.epa.gov/region06/
Telephone: (214) 665-2200

Region 7: Iowa, Kansas, Missouri, Nebraska

Environmental Protection Agency
901 N. 5 Street
Kansas City, KS 66101
http://www.epa.gov/region07/
Telephone: (913) 551-7003

Region 8: Colorado, Montana, North Dakota, South Dakota, Utah, Wyoming

Environmental Protection Agency
999 18th Street, Suite 500
Denver, CO 80202-2466
http://www.epa.gov/region08/
Telephone: (303) 312-6312

Region 9: Arizona, California, Hawaii, Nevada

Environmental Protection Agency
75 Hawthorne Street
San Francisco, CA 94105
http://www.epa.gov/region09/
Telephone: (415) 947-8000

Region 10: Alaska, Idaho, Oregon, Washington

Environmental Protection Agency
1200 Sixth Avenue
Seattle, WA 98101
http://www.epa.gov/region10/
Telephone: (206) 553-1200

Test Yourself

Questions

1. True or false? "An open container of refrigerant will not boil at room temperature."

2. What can happen to the skin if you spill refrigerant on it?

3. What can happen if you heat up a small, closed container of refrigerant?

4. Why is moisture harmful to the air conditioning system?

5. What service tool do you use to remove excess moisture from the air conditioning system?

6. True or false? "Use only a good grade of motor oil in the air conditioning system."

7. What is the lubricant used with Refrigerant-134a?

See "*Answers to Chapter 2 Questions*" on page B-1.

Basic System: How It Works

Introduction

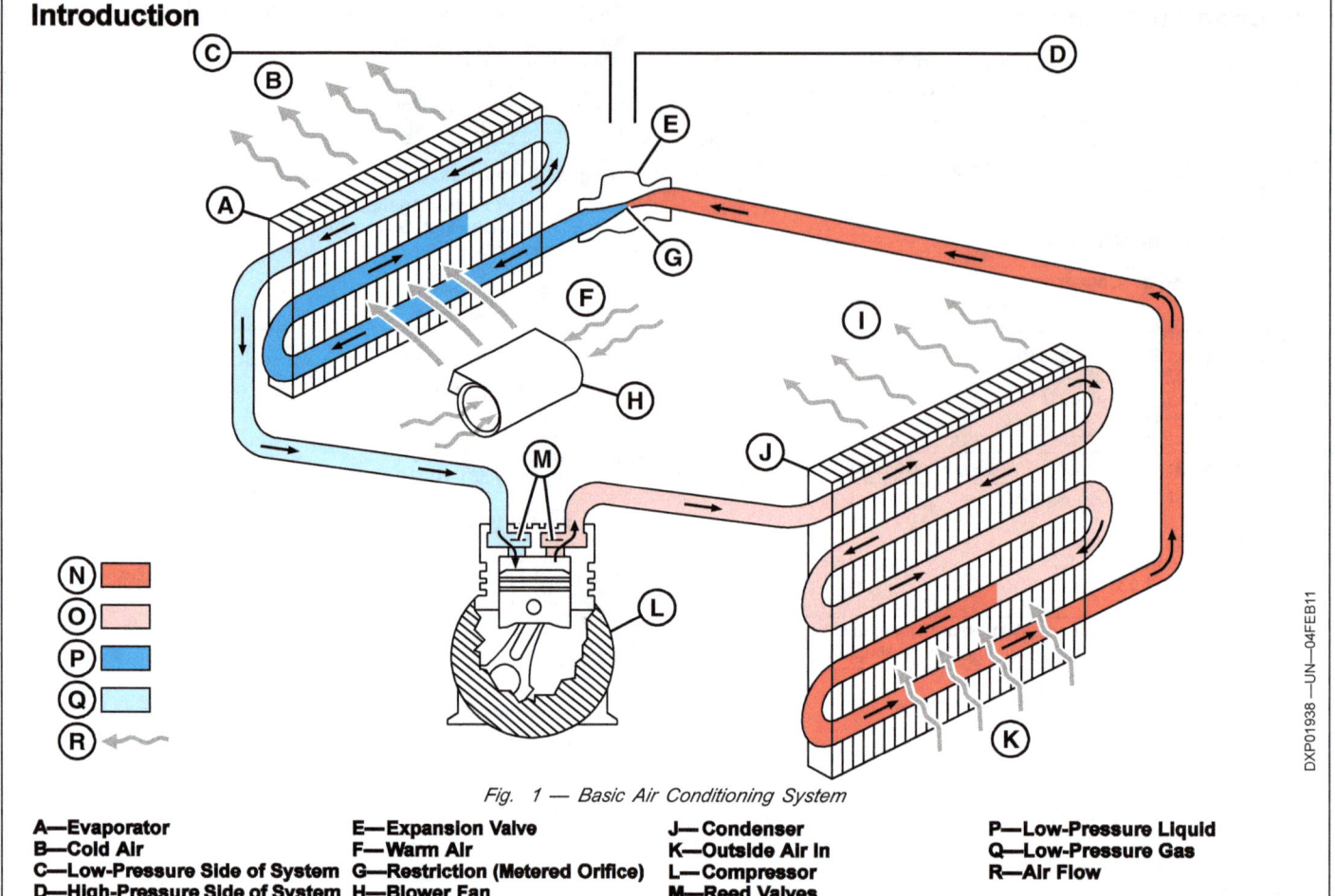

Fig. 1 — Basic Air Conditioning System

A—Evaporator
B—Cold Air
C—Low-Pressure Side of System
D—High-Pressure Side of System
E—Expansion Valve
F—Warm Air
G—Restriction (Metered Orifice)
H—Blower Fan
I—Hot Air Out
J—Condenser
K—Outside Air In
L—Compressor
M—Reed Valves
N—High-Pressure Liquid
O—High-Pressure Gas
P—Low-Pressure Liquid
Q—Low-Pressure Gas
R—Air Flow

NOTE: *An animated display of how the components in a basic Air Conditioning System works is available on the Instructor Art CD.*

All air conditioning systems must have four basic elements:

- Compressor
- Condenser
- Expansion Valve
- Evaporator

Fig. 1 shows the actual parts of the system. Note how the basic refrigerant cycle of Chapter 1, Fig. 11 is repeated.

From the **compressor**, high-pressure gas is sent to the **condenser**, where the heat is dissipated and condensed to liquid. The high-pressure liquid flows on to the **expansion valve**, where it is metered and its pressure is reduced. At the **evaporator**, the liquid absorbs heat from the air and evaporates to gas. The cycle is then repeated, starting at the compressor.

The air conditioner is a heat transfer unit. A failure of any element will interrupt the heat exchange cycle and cause the whole system to fail. Each basic element is engineered and balanced to the other parts of the system in order to move heat from the inside air to the outside air, where it is dissipated.

Let's look at each basic element of the system.

Compressor

The purpose of the **compressor** is to circulate the refrigerant in the system under **pressure**, thus concentrating the heat it contains. At the compressor the low-pressure gas is changed to high-pressure gas. This pressure buildup can only be accomplished by having a restriction in the high-pressure side of the system. This is a small valve located in the expansion valve. The metered orifice shown in Fig. 1 will serve the purpose for our basic system.

The compressor (Fig. 2 and Fig. 3) has reed valves to control the entrance and exit of refrigerant gas during the pumping operation. These must seat firmly. For instance, an improperly seated intake reed valve can result in gas leaking back into the low side during the compression stroke, thus raising the low-side pressure and impairing the cooling effect. Likewise, a badly seated discharge reed valve can allow condensing or head pressure to drop as it leaks past the valve, lowering the efficiency of the compressor.

A—Piston on downstroke.
B—Downstroke of piston creates vacuum in cylinder. Pressure in suction line forces intake reed valve open.
C—Pressure in discharge line holds discharge reed valve closed.
D—Pressure in cylinder holds intake reed valve closed.
E—Pressure in cylinder raises the discharge reed valve, and gas flows into discharge pipe.
F—Piston on upstroke.

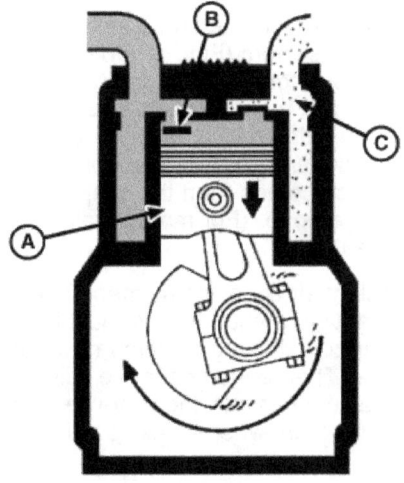

Fig. 2 — Intake Stroke

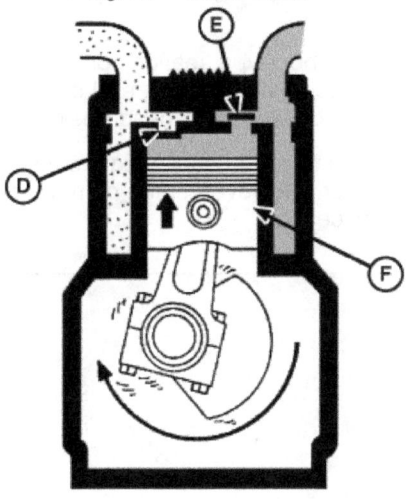

Fig. 3 — Compression Stroke

Basic System: How It Works

Two service valves are located near the compressor as an aid in servicing the system (Fig. 4). One services the high side and one is used for the low side. A fitting is provided on each for attaching the test gauge hoses for testing the system. The high-side service valve is quickly identified by the smaller discharge hose to the condenser, while the low-side, which comes from the evaporator, is larger than the discharge hose. (For more details, see "Service Valves" on page 04-6.)

Some independent air conditioning manufacturers still use service valves having a shutoff valve built in. However, many factory air conditioning systems now use only Schrader valves. The gauge hoses are still connected to the Schrader valve fitting, in which a valve is incorporated to hold in the refrigerant when a test hose is not connected to it.

The service points on the typical compressor are as follows:

- Replaceable carbon-type seal on compressor shaft
- Gasket at opposite, or oil pump, end of unit
- Serviceable reed valves and head gasket

Any other internal failures usually require the replacement of the compressor. However, some compressors are completely serviceable.

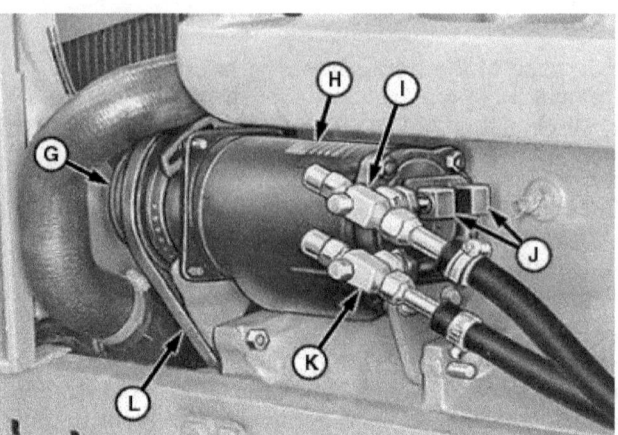

Fig. 4 — Typical Compressor (Axial Piston-Type Shown)

G—Pulley
H—Compressor
I—Low-Side Service Valve
J—Manifold
K—High-Side Service Valve
L—Compressor Drive Belt

The compressor is normally belt-driven from the engine crankshaft. Most manufacturers use a magnetic-type clutch that provides a means of stopping the pumping of the compressor when refrigeration is not desired.

Compressor Relief Valve

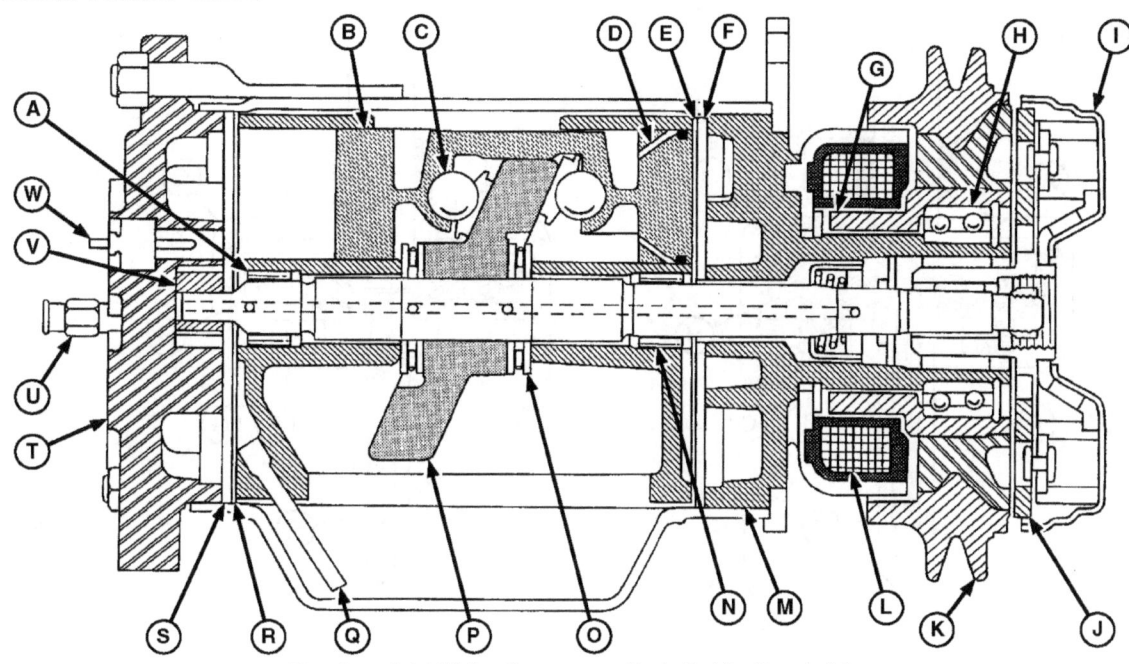

Fig. 5 — Axial Piston Compressor Controlled by Swashplate

A—Mainshaft Rear Bearing
B—Piston
C—Drive Ball
D—Suction Reed
E—Piston Ring
F—Front Discharge Valve Plate
G—Shaft Seal
H—Pulley Bearing
I—Dust Cover
J—Hub and Drive Plate Assembly
K—Pulley
L—Clutch Coil
M—Front Head
N—Mainshaft Front Bearing
O—Mainshaft Thrust Bearing
P—Swashplate
Q—Oil Pickup Tube
R—Suction Reed
S—Rear Discharge Valve Plate
T—Rear Head
U—Relief Valve
V—Oil Pump
W—Superheat Shutoff Switch

Some compressors have a **relief valve** (Fig. 5) for regulating pressure. If the system discharge pressure exceeds rated pressure, the valve will open automatically and stay open until the pressure drops. The valve will then close automatically.

If the relief valve opens, some oil may be ejected through the valve. Correct any conditions that would cause this valve to open.

Superheat Shutoff Switch

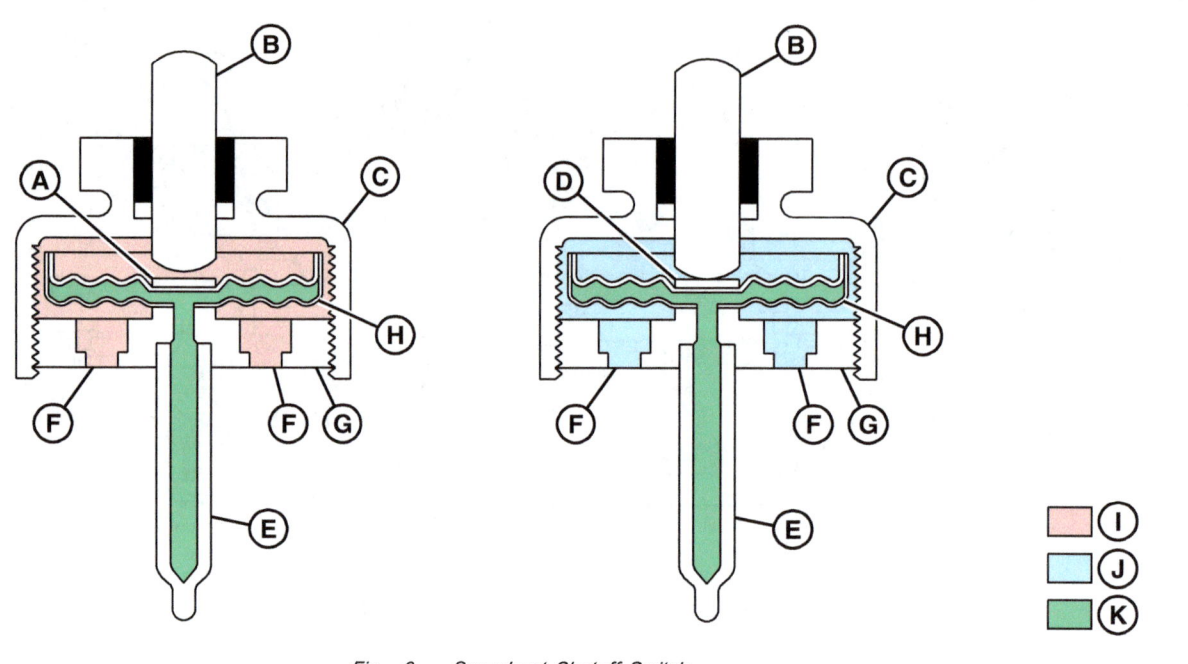

Fig. 6 — Superheat Shutoff Switch

A—Electrical Contact (Open)
B—Terminal
C—Housing
D—Electrical Contact (Closed)
E—Temperature Sensing Tube
F—Base Openings (4)
G—Mounting Base
H—Diaphragm
I—High-Pressure Gas
J—Low-Pressure Gas
K—Trapped Gas

A **superheat shutoff switch** (Fig. 6) stops the compressor when a combination of compressor suction pressure and the compressor temperature is too high. This prevents possible damage to the compressor from lack of refrigerant oil.

The switch is mounted in the rear head of the compressor. It senses refrigerant pressure in the low side of the compressor.

The superheat shutoff switch is a mechanical switch sensitive to both temperature and pressure. An electrical contact, welded to the diaphragm, contacts the terminal whenever there is low refrigerant pressure or high temperature in the compressor inlet. This stops the compressor as explained next.

Four holes in the mounting base allow gas at the inlet of the compressor to act upon the outside areas of the diaphragm. The diaphragm and sensing tube assembly are charged with a refrigerant gas. The temperature sensing tube protrudes into the suction cavity of the compressor to sense the refrigerant operating temperature. When the refrigerant charge is adequate and temperature is normal, the system pressure is high enough to keep the electrical contact away from the terminal.

If the system develops extremely low pressure (possibly because of a loss of refrigerant), the pressure necessary to overcome the expansion of the gas inside the diaphragm and sensing tube is reduced. A loss in system pressure also increases the operating temperature of the gas to the inlet of the compressor. The increase in temperature expands the gas in the sensing tube, which pushes the electrical contact against the terminal. When closed, the current flow is routed through a thermal fuse to ground. This stops the compressor action.

If the condition does not correct itself in two or three minutes, the excess heat generated by a resistor in the thermal fuse will cause the fuse link to burn out. This stops current flow to the compressor and stops the air conditioning system.

High- and Low-Pressure Switches

Newer air conditioning systems may use **high- and low-pressure switches** to protect the system rather than a superheat switch.

A high-pressure condition can result from insufficient air flow across the condenser. If this happens, the high-pressure switch (Fig. 7) opens and shuts off the compressor. Normally the switch is closed. The low-pressure switch protects the compressor if the refrigerant is lost. When the system pressure is low, the switch opens and shuts off the compressor. The switch can also activate if there is low air flow across the evaporator.

The function of the combined high- and low-pressure switch also named high/low pressure switch (Fig. 8) is the same.

A—High-Pressure Switch
B—Low-Pressure Switch
C—Combined High- and Low-Pressure Switch

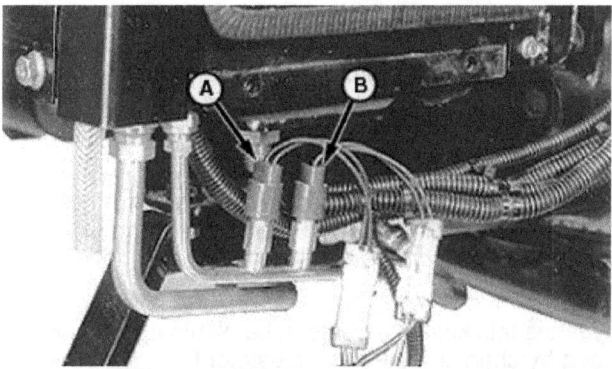

Fig. 7 — High- and Low-Pressure Switches

Fig. 8 — Combined High- and Low-Pressure Switch in a Modern Tractor

Choice of Compressor

The compressor is actually installed in the system for two purposes. The most important of these is heat concentration by compression. Second is the circulation of refrigerant through the system.

Overall dimensions and weight must be considered, plus the method used to drive the compressor. In most installations the compressor fits into a confined area where size and weight become an important factor.

The choice of compressors may be a two-piston crankshaft-controlled reciprocating piston type (Fig. 2), a multi-axial piston-swashplate controlled type (Fig. 5), a radial piston or a rotary vane compressor.

The compressor must concentrate heat molecules contained in the low-pressure refrigerant returning from the evaporator to a temperature much higher than the ambient or outside air temperature. The wide differential between the refrigerant and ambient temperatures is necessary to aid rapid heat flow in the condenser from the hot refrigerant gas to the much cooler outside air. Remember, heat will flow only from the warmer to the cooler.

Little heat is added to the refrigerant by the operation of the compressor. The heat felt on the compressor housing is caused by compression of refrigerant vapor. Some heat is lost through the walls of the compressor by radiation, which compensates for the heat resulting from friction of moving parts within the compressor.

To create high-pressure gas, the compressor must move a large volume of refrigerant vapor rapidly past its discharge valve into the high side of the system. The small orifice dividing the high side from the low side of the system provides a pressure for the compressor to pump against. Too small an orifice or a compressor with too large a capacity could cause excessive buildup of pressure on the high side of the system. Too large an orifice in the expansion valve, a compressor with too small a capacity, or a reed valve failure could prevent a buildup of pressure high enough to allow sufficient heat exchange in the condenser.

Compressor Noise Complaints

Many noise complaints can be traced to the compressor mount and drive. If a unit is noisy at one speed and quiet at another, it is not compressor noise.

Each machine has its critical frequencies where all vibrations get into harmony to generate sound or noise. The speed where these critical points are found will vary with each machine and each mount and drive arrangement.

Many times this kind of noise can be eliminated or greatly reduced by changing the belt adjustment.

Noise doesn't necessarily mean compressor replacement. Usually tightening mounts, adding idlers, or changing belt adjustment and length are more successful in removing or reducing this type of noise than replacing the compressor.

Noises from the clutch are difficult to recognize because the clutch is so close to the compressor. A loose bolt holding the clutch to the shaft will make a lot of noise.

The difference between suction pressure and discharge pressure also plays an important part in sound level.

A compressor with low suction pressure will be noisier than one with a higher pressure. Likewise, high head pressures tend to make compressors noisy because of the extra load on bearings.

Also, consider whether the system is properly charged, whether the expansion valve is feeding properly to use the evaporator efficiently, and whether enough air is being fed over the evaporator coil.

On the high side, check for contaminants in the system as well as cleanliness of condenser, amount of air that can flow through it, and overcharge of refrigerant.

Since a compressor has many moving parts, it is normal for it to generate some noise just as a motor generates some noise as it operates. The refrigerant gases will also produce noises and vibrations as they are moved by the compressor pistons.

The level of permissible noise varies with each customer. Generally an explanation of how the compressor functions and what it does will satisfy a customer: since the compressor is doing the work, it is natural for it to make some sound.

Condenser

The purpose of the **condenser** (Fig. 9) is to receive the high-pressure gas from the compressor and convert this gas to a liquid. It does this by heat transfer, or the principle that heat will always move from a warmer to a cooler substance. Air passing over the condenser coils carries off the heat, and the gas condenses. The condenser often looks like an engine radiator.

Condensers used on R-12 and R-134a systems are not interchangeable. R-134a has a different molecular structure and requires a large capacity condenser.

As the compressor subjects the gas to increased pressure, the heat intensity of the refrigerant is actually concentrated into a smaller area. This raises the temperature of the refrigerant higher than the ambient temperature of the air passing over the condenser coils. Clogged condenser fins will result in poor condensing action and decreased efficiency.

A factor often overlooked is flooding of the condenser coils with refrigerant oil. Flooding results from adding too much oil to the system. Oil flooding is characterized by poor condensing, resulting in increased head pressure

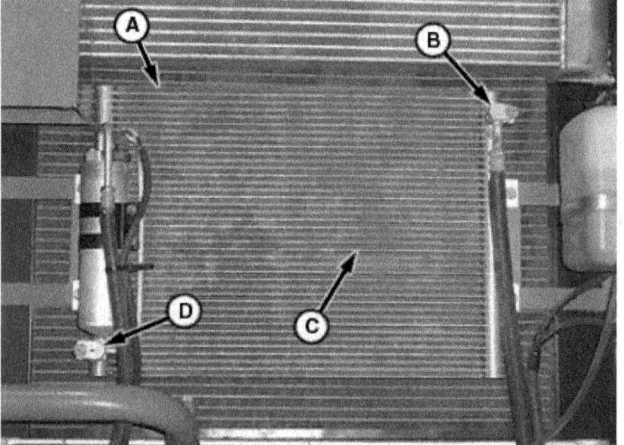

Fig. 9 — Condenser

A—Core
B—Inlet
C—Fins to Dissipate Heat
D—Outlet

and pressure on the low side. This combination results in poor cooling from the evaporator.

Basic System: How It Works

Types of Condensers

Heat exchange in the condenser is accomplished by two types of condensers, air-cooled and water-cooled. Water-cooled condensers are not used in the mobile machine field and will not be considered here.

There are two basic types of air-cooled condensers:

- Ram Air — for automotive systems
- Forced Air — for off-road machines

Ram air condensers depend upon vehicle movement to force a large volume of air past the fins and coils of the condenser. The engine fan is used to increase air volume at lower speeds. The clutch-type fan is designed to allow the fan blades to free-wheel at higher speeds to eliminate blade drag. At lower speeds, the fan clutch will engage the fan to increase air flow over the condenser and radiator coils.

Forced air condensers use electric fans to move a large volume of air over the condenser (Fig. 10). Farm and industrial machines must rely on forced air to remove heat from the refrigerant. This is because these machines must operate for long periods at slow speeds, which requires that air be forced over the condenser for proper cooling.

Condensing action is the change of state of the refrigerant from a vapor to a liquid. It is controlled by:

- Pressure of the refrigerant
- Air flow over the condenser

Condensing pressure is the controlled pressure of the refrigerant as it condenses to a liquid.

Normal Condenser Pressure

Condensing pressure must be high enough to create a wide temperature differential between the heat-laden refrigerant and the hot air passing over the condenser fins and coils. Only in this way can the air carry off enough heat for proper cooling.

Too-High Condenser Pressure

Fig. 10 — Forced Air Condenser

A—Condenser Core   B—Condenser Fan

Indicated by: Excessive head pressure on high-side gauge.

Caused by: Restriction of refrigerant flow in high side of system or lack of air flow over condenser coils.

Poor condensing action results in the too-high pressure. The upper two-thirds of the condenser coils remove heat from the hot refrigerant vapor, while the lower third contains liquid. Too high a condensing pressure will upset this balance, allowing superheated vapor to enter the liquid hose and the expansion valve.

Too-Low Condenser Pressure

Indicated by: Higher than normal pressure on low-side gauge.

Caused by: Failed compressor reed valve or piston. Heat exchange in the condenser will be cut down, and the excessive heat will remain in the low side of the system.

Expansion Valve

The **expansion valve** removes pressure from the liquid refrigerant to allow expansion or change of state from a liquid to a vapor in the evaporator.

The high-pressure liquid refrigerant entering the expansion valve is quite warm. This may be verified by feeling the liquid line at its connection to the expansion valve. The liquid refrigerant leaving the expansion valve is quite cold. The orifice within the valve does not remove heat, but only reduces pressure. Heat molecules contained in the liquid refrigerant are thus allowed to spread as the refrigerant moves out of the orifice. Under a greatly reduced pressure the liquid refrigerant is at its coldest as it leaves the expansion valve and enters the evaporator.

Pressures at the inlet and outlet of the expansion valve will closely approximate gauge pressures at the inlet and outlet of the compressor in most systems. The similarity of pressures is caused by the closeness of the components to each other. Slight variation in pressure readings of a very few pounds is due to resistance, causing a small pressure drop in the lines and coils of the evaporator and condenser.

Three types of valves are used on machine air conditioning systems:

- Sealed sending bulb expansion valve
- Internally equalized valve — most common
- Externally equalized valve — special control

Let's look at each valve in detail.

Sealed Sensing Bulb Expansion Valve

In a **sealed sensing bulb expansion valve** (Fig. 11), the valve diaphragm is activated by sensing temperature and pressure within the valve body. The sealed sensing bulb senses the evaporator outlet discharge temperature and pressure of the refrigerant as it passes through the valve back to the low-pressure or suction side of the compressor.

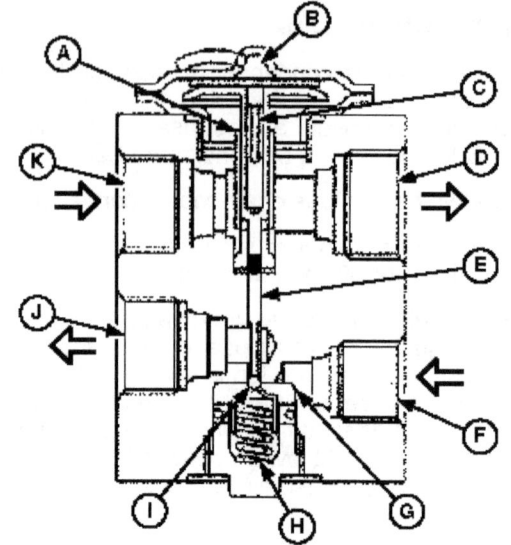

Fig. 11 — Sealed Sensing Bulb Expansion Valve

A—Internal Equalization Passage
B—Valve Diaphragm
C—Sealed Sensing Bulb
D—Outlet (to Compressor)
E—Operating Pin
F—Inlet (from Condenser)
G—Metering Orifice
H—Ball Seat
I—Ball Seat Area
J—Outlet (to Evaporator)
K—Inlet (from Evaporator Discharge)

The metering orifice and ball seat area in the expansion valve is relatively small. The rapidly expanding refrigerant passing through this area can cause any moisture in the system to freeze at this point and block refrigerant flow. Other contaminants in the system can also cause a valve to malfunction. If an expansion valve malfunctions, it must be replaced because expansion valves are not repairable.

Continued on next page

Basic System: How It Works

Internally Equalized Expansion Valve

In an **internally equalized valve** (Fig. 12), the refrigerant enters the inlet and screen as a high-pressure liquid. The refrigerant flow is restricted by a metered orifice through which it must pass. As the refrigerant passes through this orifice, it changes from a high-pressure liquid to a low-pressure liquid (or passes from the high side to the low side of the system).

Let's review briefly what happens to the refrigerant as we change its pressure.

As a high-pressure liquid, the boiling point of the refrigerant has been raised in direct proportion to its pressure. This has concentrated its heat content into a small area, raising the temperature of the refrigerant higher than that of the air passing over the condenser. This heat will then transfer from the warmer refrigerant to the cooler air, which condenses the refrigerant to a liquid.

The heat transferred into the air is called latent heat of condensation. Four pounds (1.8 kg) of refrigerant flowing per minute through the orifice will result in 12,000 Btu (12.7 MJ) per hour transferred, which is designated as a one-ton unit. Six pounds (2.7 kg) of flow per minute will result in 18,000 Btu (19.0 MJ) per hour, or a one-and-one-half ton unit.

The refrigerant flows through the metered orifice is of vital importance. Anything that restricts this flow seriously affects the operation of the whole system.

For example, a one-ton system uses an orifice approximately 0.080 inch (2 mm) in diameter during maximum cooling. A restriction of this orifice can prevent the unit from cooling to its full capabilities.

If the area cooled by the evaporator suddenly gets colder, the heat transfer requirements of the evaporator are changed. If the expansion valve kept feeding the same

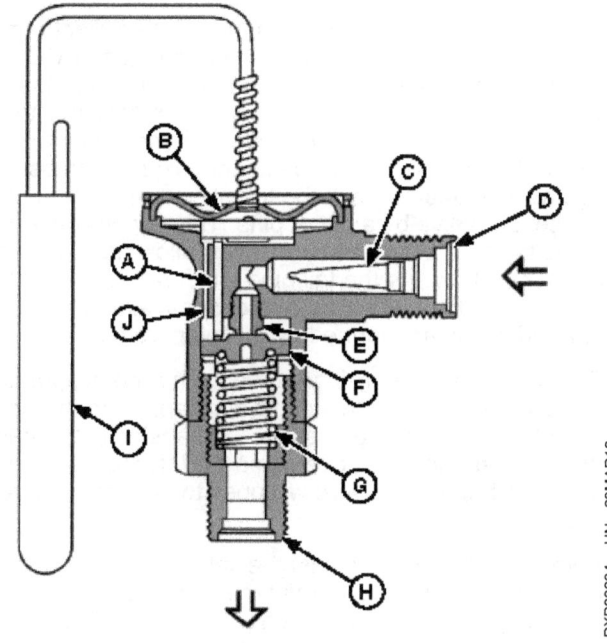

Fig. 12 — Internally Equalized Valve

A—Pin
B—Diaphragm
C—Screen
D—Inlet (from Condenser)
E—Orifice
F—Valve Seat
G—Superheat Spring
H—Outlet (to Evaporator)
I—Thermal Sensing Bulb
J—Internal Equalizing Passage

amount of refrigerant to the evaporator, the fins and coils would be cooled until they froze over with ice and stopped all air flow.

To control this, the orifice in the expansion valve is metered to vary the flow of refrigerant. This is done by an internal balance of pressures, which moves the valve seat.

Continued on next page

A **thermal bulb** (Fig. 13) is connected to a diaphragm by a swell line and filled with refrigerant gas or CO_2 (carbon dioxide) is secured firmly by a clamp to the evaporator tail pipe. The thermal bulb is sensitive to tail pipe temperatures. If the tail pipe becomes warm, the gas inside the bulb will begin to expand, exerting pressure against the diaphragm in the top plate that is connected to the seat or valve by a pin or pins. This expansion will then move the seat away from the orifice, allowing an increased refrigerant flow. As the tail pipe temperature drops, the pressure in the thermal bulb also drops, allowing the valve to restrict flow as required by the evaporator.

The pressure of the refrigerant entering the evaporator is fed back to the underside of the diaphragm through the internal equalizing passage. Expansion of the gas in the thermal bulb must overcome the internal balancing pressure before the valve will open to increase refrigerant flow.

A spring is installed against the valve and adjusted to a predetermined setting at the time of manufacture. This is the superheat spring, which prevents flooding of the evaporator with excessive liquid.

Superheat is an increase in temperature of the gaseous refrigerant above the temperature at which the refrigerant vaporizes. The expansion valve is designed so that the temperature of the refrigerant at the evaporator outlet must have 8–12°F (4–7°C) of superheat before more refrigerant is allowed to enter the evaporator.

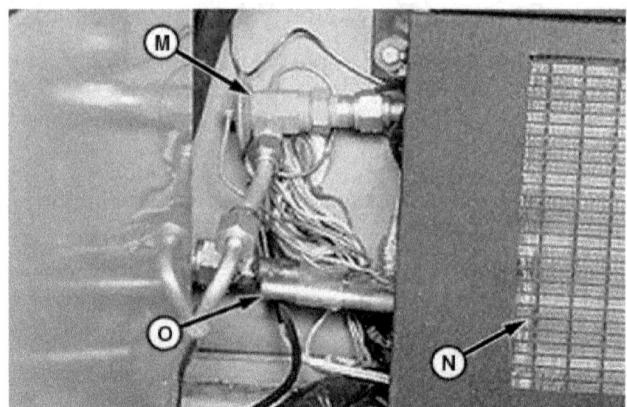

Fig. 13 — Expansion Valve with Thermal Bulb

M—Expansion Valve Body   **O—Thermal Bulb**
N—Evaporator

The adjusted tension of this spring is the determining factor in the opening and closing of the expansion valve. During opening or closing, the spring tension retards or assists valve operation as required.

Normally, this spring is never adjusted in the field. Tension is adjusted from 4–16° as required for the unit on which it is to be installed. This original setting is sufficient for the life of the valve, and special equipment is required in most cases to accurately calibrate this adjustment.

Externally Equalized Expansion Valve

Operation of the **externally equalized valve** (Fig. 14) is the same as that of the internal type except that evaporator pressure is fed against the underside of the diaphragm from the tail pipe of the evaporator by an equalizer line. This balances the temperature of the tail pipe through the expansion valve thermal bulb against the evaporator pressure taken from the tail pipe.

A—External Equalizer Line   **C—Outlet (to Evaporator)**
B—Inlet (from Condenser)

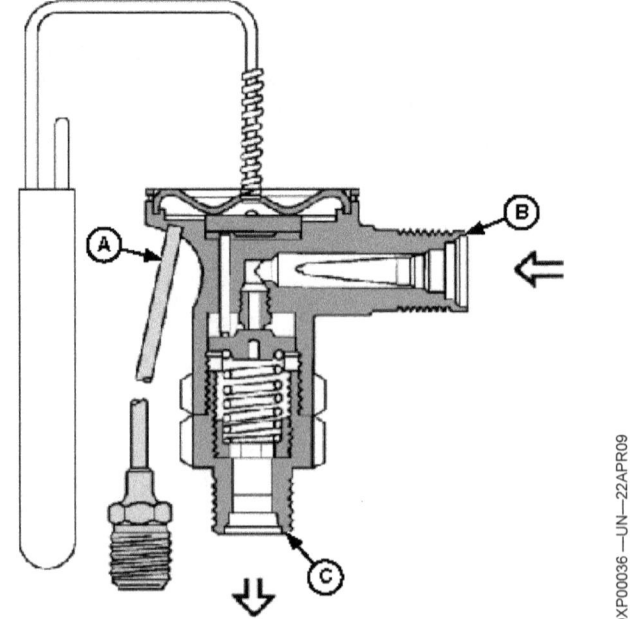

Fig. 14 — Externally Equalized Valve

Service Precautions

The thermostatic expansion valve is more sensitive to foreign materials than any other unit in the air conditioning system.

Observe the following rules to protect the valve:

1. In winter, advise the customer to turn over the pulley of the compressor by hand several times periodically to prevent the internal moving parts from corroding and sticking. This will also lubricate the compressor seal, to properly seal the refrigerant. Normally, do not operate the system in winter because this can cause flooding of the compressor.

2. When servicing the system, clean or replace all accessible screens.

3. Install a filter if the system has excessive foreign materials.

4. Evacuate the system to remove moisture from the system.

5. Cap or cover lines opened for service to prevent entry of moisture or dirt.

6. Replace the receiver-dryer or accumulator as soon as excess moisture content is evident. Any system should have the receiver-dryer replaced at least upon opening the system the third time for service. Also, anytime the system has been opened for a long period due to an accident or rupture, replace the receiver-dryer.

7. Handle the thermal bulb and line with extreme care; excessive bending and rough handling can cause a break that will release the gas, ruining the valve.

8. Use a back-up wrench when removing any connection to prevent twisting of a line, which may result in weakening and breaking it.

9. Replace the expansion valve only with a comparable valve. A numbering system is used to designate the orifice size. Replacing the valve with one that has too small or too large an orifice can seriously affect system operation.

10. Maintain a positive contact between the thermal bulb and tail pipe. Over a period of time, corrosion will form between the two contact surfaces and insulate them so that the operation of the expansion valve will be affected. The clamp may also work loose, preventing positive contact between the two.

Evaporator

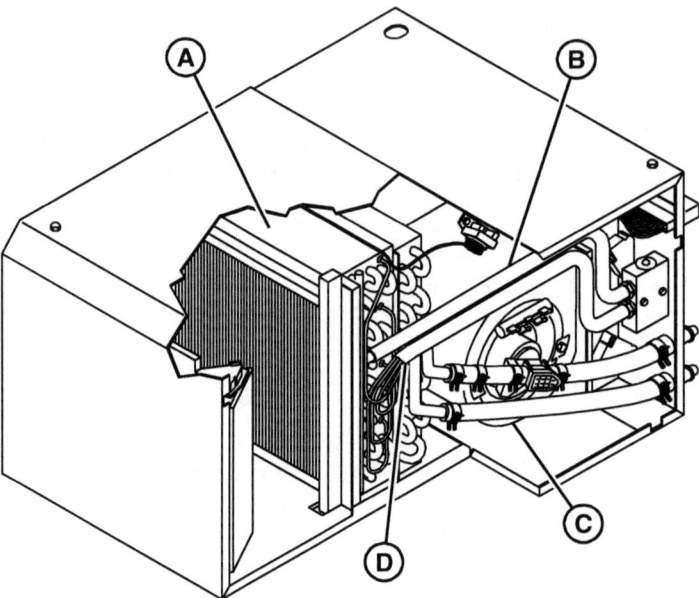

Fig. 15 — Evaporator

A—Evaporator Core
B—Inlet
C—Blower Fan
D—Outlet

The **evaporator** (Fig. 15) works the opposite of the condenser, for here refrigerant liquid is converted to gas, absorbing heat from the air in the compartment.

When the liquid refrigerant reaches the evaporator, its pressure has been reduced, dissipating its heat content and making it much cooler than the fan air flowing around it. This causes the refrigerant to absorb heat from the warm air and reach its low boiling point rapidly. The refrigerant then vaporizes, absorbing the maximum amount of heat.

This heat is then carried by the refrigerant from the evaporator as a low-pressure gas through a hose or line to the low side of the compressor, where the whole refrigeration cycle is repeated.

The evaporator removes heat from the area that is to be cooled. The desired temperature of cooling of the area will determine if refrigeration or air conditioning is desired. For example, food preservation generally requires low refrigeration temperatures, ranging from 40°F (4°C) to below 0°F (−18°C).

A higher temperature is required for human comfort. A larger area is cooled, which requires that large volumes of air be passed through the evaporator coil for heat exchange. A blower becomes a necessary part of the evaporator in the air conditioning system. The blower fans must not only draw heat-laden air into the evaporator, but must also force this air over the evaporator fins and coils — where it surrenders its heat to the refrigerant — and then forces the cooled air out of the evaporator into the space being cooled.

Fan Speed

Fan speed is essential to the evaporation process in the system. Heat exchange, as we explained under condenser operation, depends upon a temperature differential of the air and the refrigerant. The greater the differential, the greater the amount of heat exchanged between the air and the refrigerant. A high heat load, as is generally encountered when the system is turned on, will allow rapid heat transfer between the air and the cooler refrigerant.

The blower fan turned on to its highest speed will deliver its greatest volume of air across the fins and coils for rapid evaporation. As the area is cooled, it will soon reach a temperature where little extra cooling will result if the fan is allowed to continue its high-volume flow. A reduction in fan speed will decrease volume, but the lower volume rate will allow the air to remain in contact with the fins and coils for a longer period of time and surrender its heat to the refrigerant.

Both condensing and evaporating processes depend upon wide temperature differentials for rapid heat exchange. A lowering of temperature of the refrigerant in the condenser will affect the condensing process, and a lowering of the temperature of the air under air conditioning will slow down the evaporating process.

Cooling the evaporator is dependent on controlled air flow over the evaporator coils by regulating the fan blower speed.

NOTE: For the coldest air temperature from the evaporator, operate the blower fan at the lowest speed possible to allow the greatest heat absorption by the refrigerant from the air.

Problems of Flooded or Starved Evaporator Coils

NOTE: *An animated display of how the components in a basic Air Conditioning System works is available on the Instructor Art CD.*

Changing the state of the refrigerant in the evaporator coils is as important as the air flow over the coils. Liquid refrigerant supplied to the coils by the expansion valve expands to a vapor as it absorbs heat from the air. Some liquid refrigerant must be supplied throughout the total length of the evaporator coils for full capacity.

A **starved** evaporator coil is a condition in which not enough refrigerant has been supplied through the total coil length. Therefore, expansion of the refrigerant has not occurred through the whole coil length, resulting in poor coil operation and too-low heat exchange.

A **flooded** evaporator is the opposite of the starved coil. Too much refrigerant is passed through the evaporator coils, resulting in unexpanded liquid passing into the suction line and into the compressor. Liquid refrigerant in the compressor can result in damage to the reed valves and pistons. A flooded evaporator will contain too much refrigerant for efficient heat absorption in the evaporator coil. The result is lack of evaporation and so poor evaporator cooling.

Gauge pressure readings on the low side of the system readily indicate either condition:

A **starved coil** is shown by too low a reading on the compound gauge plus a too-quick frost formation on the fins. Also, too little air is emitted from the evaporator.

A **flooded coil** is indicated by too high a pressure on the compound gauge and excessive sweating of the evaporator coils and suction hose. This is accompanied by little cooling from the evaporator.

The basic system we have discussed will work okay under constant loads or until the unit ices up because of too much humidity. The cooling rate of the evaporator can be controlled to a great extent by varying the speed of the fan.

However, adding other features will aid in operation of the system. Let's look at several of the extra controls, starting with the receiver-dryer (dehydrator) shown in Fig. 16.

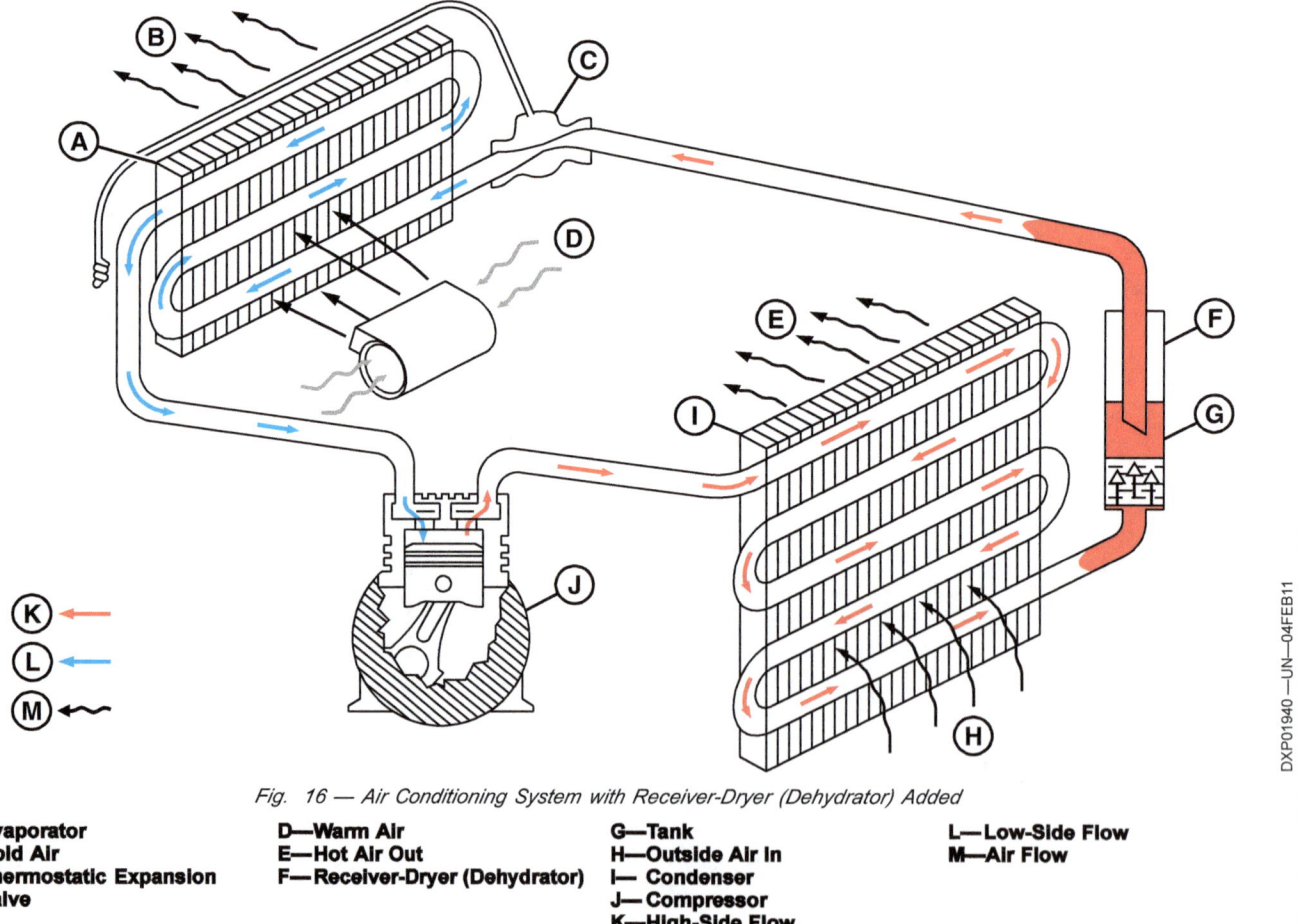

Fig. 16 — Air Conditioning System with Receiver-Dryer (Dehydrator) Added

A—Evaporator
B—Cold Air
C—Thermostatic Expansion Valve
D—Warm Air
E—Hot Air Out
F—Receiver-Dryer (Dehydrator)
G—Tank
H—Outside Air In
I—Condenser
J—Compressor
K—High-Side Flow
L—Low-Side Flow
M—Air Flow

Receiver-Dryer (Dehydrator)

Air conditioning systems do not operate at 100% efficiency. Over a period of time, very slight leaks and occasionally some serious leaks will develop. In addition, the demand for refrigerant by the evaporator varies with changes in the heat load, the condensing action, and pump speed.

To compensate for these variables, a **receiver-dryer** is provided in the system (Fig. 16). Here the refrigerant is stored until needed by the evaporator. The addition of the receiver tank increases the capacity of the system approximately one to one and a half pounds of refrigerant.

A **desiccant** or drying agent such as silica gel or molecular sieve in R-12 systems or zeolite in R-134a systems is sealed inside the receiver during its manufacture (Fig. 17).

If the air conditioning system is operated during high outside temperatures, the dryer will hold back approximately 50 drops of moisture from circulating in the system. If the desiccant has reached its saturation point, the excess moisture will circulate in the system. Operation below approximately 80°F (27°C) ambient temperature will cause the desiccant to release more moisture into the system. These moisture droplets will collect inside the tubing at the inlet or outlet of the evaporator and change to ice. This moisture turning to ice will restrict the refrigerant flow and decrease cooling action of the evaporator.

Regardless of where the receiver-dryer is located, the desiccant can absorb and hold only a predetermined amount of moisture. If operation of the system is satisfactory during temperatures above 80°F (27°C)[1], but cooling action becomes intermittent below 80°F (27°C), the desiccant is super-saturated and the moisture freezes in the evaporator. As soon as the evaporator warms sufficiently to melt the ice, the flow of refrigerant will resume until more icing occurs. Above 80°F (27°C) moisture will not freeze in the system because the temperature of the refrigerant coming out of the expansion valve will be above 32°F (0°C) — the freezing point of water.

IMPORTANT: A refrigeration system cannot tolerate moisture because of the acids produced by the combination of moisture and refrigerant. These acids cause internal system corrosion.

Separate receiver-dryer or accumulator units are available to be installed in the liquid line in addition to the receiver-dryer in the system, and these will assist in removing moisture.

Some receiver-dryers contain a wet/dry indicator. Blue indicates dryer is dry. Pink indicates moisture in the

[1] Depending upon variations in operating conditions and system components, the temperature may vary from 80°F (27°C).

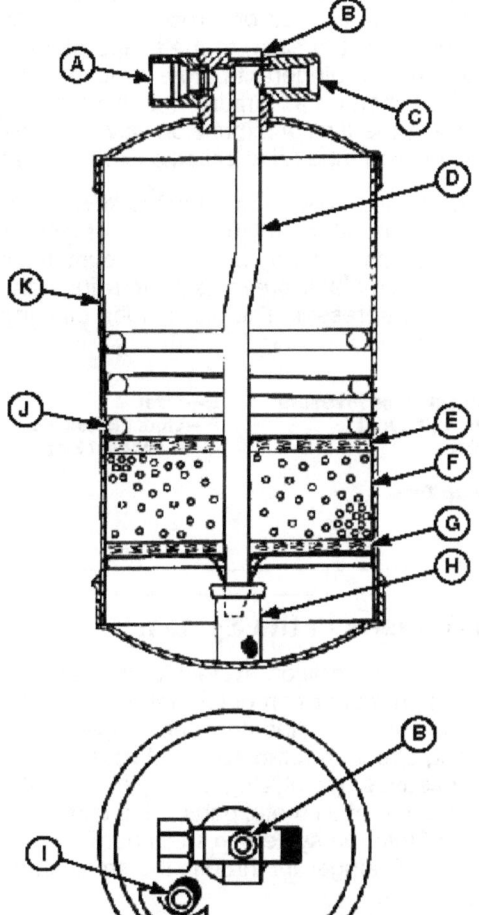

Fig. 17 — Receiver-Dryer (Dehydrator)

A—Inlet Port
B—Sight Glass
C—Outlet Port
D—Pickup Tube
E—Filter
F—Desiccant
G—Filter
H—Strainer
I—Wet/Dry Indicator
J—Spring
K—Receiver-Dryer

desiccant. Evacuating the system will not remove moisture from the desiccant. You must replace the receiver-dryer.

Accumulators

Some systems use an accumulator for temporary refrigerant storage instead of a receiver-dryer. An accumulator is located close to the evaporator outlet and stores excess refrigerant before it moves on to the compressor. When an accumulator is used instead of a receiver-dryer, the typical expansion valve is replaced with an expansion tube (also called a fixed-orifice tube).

An expansion tube allows refrigerant to flow continuously through the evaporator. At times all the refrigerant does not change to gas and may enter the accumulator as a liquid. The accumulator prevents liquid refrigerant from going to the compressor. The pickup tube opening inside an accumulator is shown in Fig. 18.

A—Connector for Pressure Switch (If used)
B—Outlet
C—Inlet
D—Pickup Tube
E—Filter
F—Liquid Bleed Hole
G—Desiccant Bag

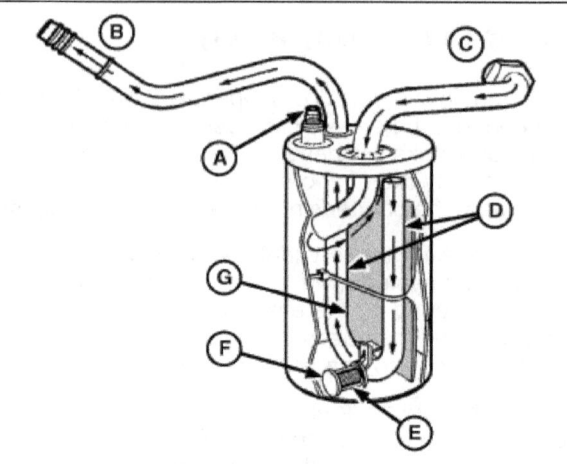

Fig. 18 — Accumulator

Use of Screens in the System

At any time a refrigeration system is opened for service, foreign matter can enter in the form of dirt or moisture. These are called noncondensibles and have a deteriorating effect on refrigerants. Moisture mixed with refrigerant causes hydrolyzing action, which results in interior corrosion of all metal parts. This corrosion will, in time, sluff off into the system in small particles that can stop the flow of refrigerant through the small orifice in the expansion valve.

Screens are installed throughout the system to filter and hold these foreign particles from circulating in the system. A filter screen is always located in the receiver-dryer (Fig. 17). Should any of these screens collect foreign particles until they can no longer pass refrigerant, refrigerant flow will stop at this point.

NOTE: Frost will form at the point of blockage of the refrigerant.

A screen can be located at the inlet of the expansion valve. Two or three manufacturers have eliminated the screen at this point in their later units. In any system containing a screen at this location it is not advisable to eliminate this screen unless a self-contained filter unit is placed ahead of the expansion valve somewhere in the liquid line between the valve and the receiver. These filters are available to be installed as a separate item.

Many manufacturers install an additional screen at the suction side or low-side service valve of the compressor. Most factory-installed units have screens located at this point.

All screens except the one located in the receiver-dryer may be removed for cleaning and should be replaced if torn or so corroded that cleaning fails to open the fine mesh.

What Happens When Refrigerant Is Blocked

A restriction or stoppage of refrigerant flow will cause the following:

- Normal or low head pressure with low suction pressure.
- Excessive coolness or frosting of the receiver-dryer or accumulator, expansion valve, and compressor service valve.
- Little or no cooling from the evaporator.

Basic System: How It Works

Thermostat and Magnetic Clutch Systems

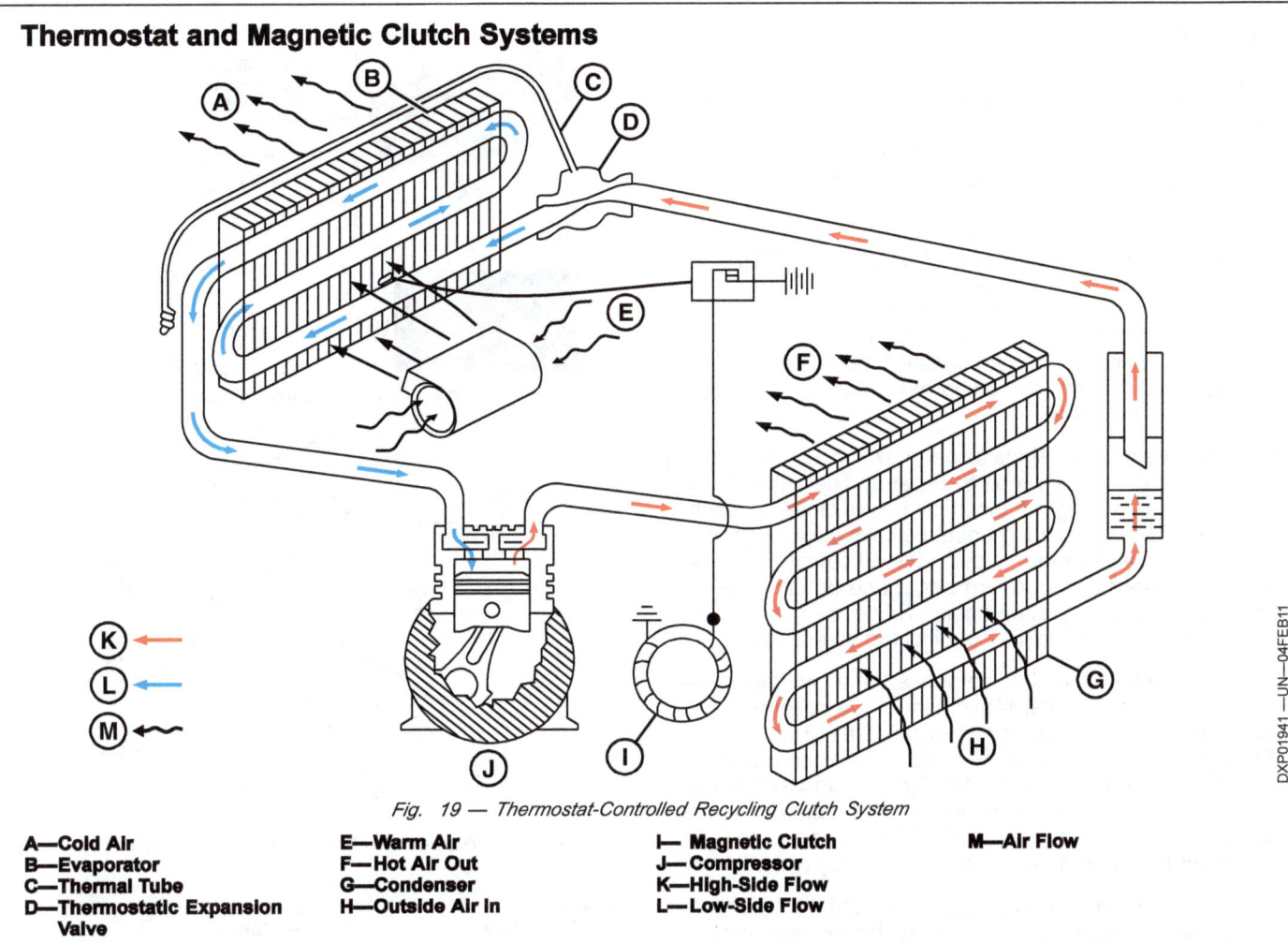

Fig. 19 — Thermostat-Controlled Recycling Clutch System

A—Cold Air
B—Evaporator
C—Thermal Tube
D—Thermostatic Expansion Valve
E—Warm Air
F—Hot Air Out
G—Condenser
H—Outside Air In
I—Magnetic Clutch
J—Compressor
K—High-Side Flow
L—Low-Side Flow
M—Air Flow

NOTE: An animated display of how the components in a basic Air Conditioning System works is available on the Instructor Art CD.

During the earlier years of vehicle air conditioning, many systems did not provide a means for stopping the pumping action of the compressor. A solid pulley was installed on the compressor crankshaft, which resulted in compressor operation anytime the engine was operating. The only time the compressor could be stopped was when the belt was removed. Even with the air conditioning controls in the OFF position during cold weather operation, a slight amount of cold air would be given off by the evaporator.

Today, manufacturers are turning more and more to the **thermostat**-controlled system with a **magnetic clutch** (Fig. 19).

Thermostat Control

The opening and closing of electrical contacts in the **thermostat** (Fig. 20) are controlled by a movement of a temperature-sensitive diaphragm or bellows. The bellows has a capillary tube connected to it, which has been filled with refrigerant. The capillary tube is positioned so that it either may have the cold air from the evaporator pass over it or may be connected to the tail pipe of the evaporator. In either position, evaporator temperature will affect the temperature-sensitive compound in the capillary tube by causing it to contract as the evaporator becomes colder. The contraction of the gas will cause the bellows to contract. This separates the electrical points and breaks the electrical circuit to the compressor clutch, which stops compressor operation.

Now the evaporator begins to warm, which, in turn, causes the gas in the capillary tube to expand. The bellows will also expand, moving the electrical points closer to each other. At a predetermined point, bellows expansion will bring the points together, closing the electrical circuit to the compressor clutch, energizing it and bringing the compressor into operation again. This cycling action will be repeated as long as air conditioning is required.

The thermostatic switch is composed of a pivoting frame attached to the bellows. Movement of the bellows during its contraction and expansion will cause the frame to pivot. Springs are used to counteract and control the movement of the pivoting frame. One half of the electrical contacts are connected to the frame and the other half are mounted solidly to the body of the switch. The contacts are insulated from the metal parts to which they are attached.

The distance the contacts must travel and the spring pressure that must be overcome by the expansion of the gas in the capillary tube acting on the bellows are the factors determining at what degree of evaporator temperature the contacts will close to complete the electrical circuit to the clutch.

Most thermostats have provisions for regulating the range between opening and closing of points. Some models have a removable cover under which an adjusting screw is located. If a set screw is not found here, assume that the thermostat is non-adjustable.

In all thermostats, the spring tension and point spacing may be varied by the operator to regulate evaporator

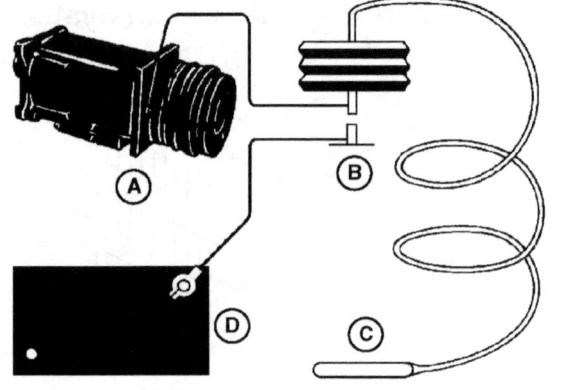

Fig. 20 — Thermostat Control

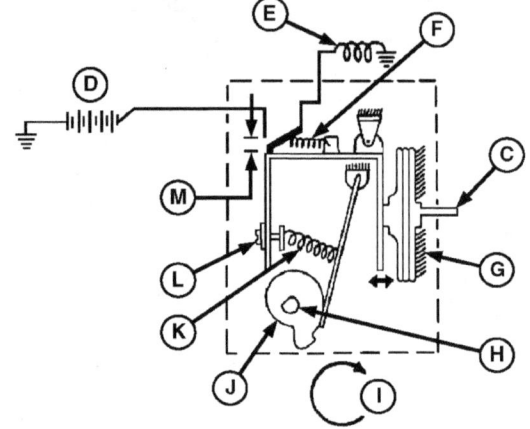

Fig. 21 — Thermostat (Thermostatic Switch Shown OFF)

A—Clutch
B—Thermostat
C—Capillary Tube
D—Battery
E—Clutch Coil
F—Over-Center Spring
G—Capillary Bellows Assembly
H—Shaft
I—Colder
J—Cam
K—Range Spring
L—Temperature Adjusting Screw
M—Point Opening

cooling for comfort. Temperature is controlled by rotating a cam (via an adjusting screw or knob control) that increases or decreases spring tension of a pivoting point (Fig. 21).

Magnetic Clutch

The clutches on machine air conditioning systems are of two types:

- Rotating coil
- Stationary coil

Rotating coil clutches have the magnetic coil inside the pulley and rotating with it. The electric current is carried to the coil by brushes mounted on the compressor frame and contacting a slip ring mounted on the inside of the rotating pulley.

Stationary coil clutches have the magnetic coil mounted on the frame of the compressor and it does not rotate. Since the coil is stationary, correct spacing is important to prevent the rotating pulley from contacting the coil, while still bringing the hub and armature into position for the fullest attraction of the magnetic force.

Each clutch manufacturer has units to fit all models of compressors according to the requirements for both six- and twelve-volt applications. The service technician replacing either the clutch unit or the coil must note carefully that the voltage of the replacement unit is correct for the vehicle on which it is to be installed.

All clutches operate on the same principle whether the magnetic coil rotates or is stationary. Each has a wound core located within a metal cup acting like a horseshoe magnet when the coil is energized electrically (Fig. 22).

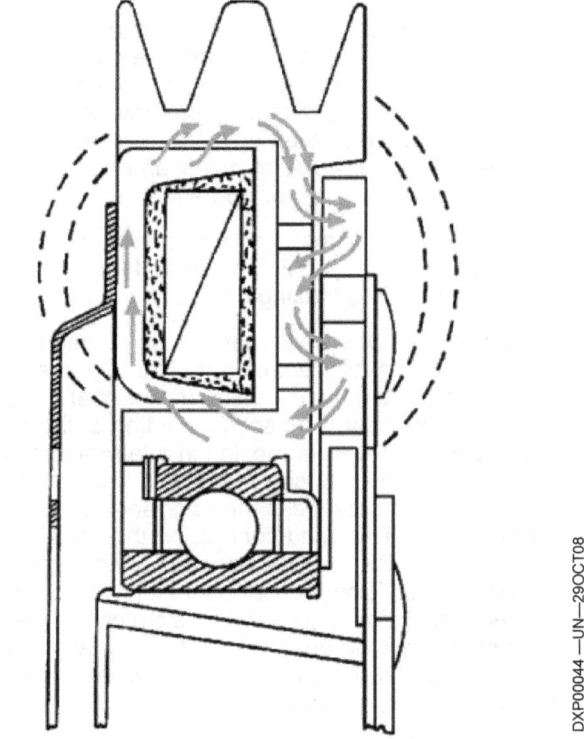

Fig. 22 — Magnetic Clutch Showing Path of Magnetic Flow

Continued on next page

Basic System: How It Works

The pulley rotates on a bearing mounted on the clutch hub (Fig. 23), except for the Frigidaire compressor, which mounts the bearing on the compressor front head assembly. The pulley is free to rotate without turning the compressor crankshaft anytime the clutch coil is not energized. The free-rotating pulley and non-energized clutch coil stop compressor operation.

An armature plate is mounted by a hub to the compressor crankshaft and is keyed into place and locked securely with a lock nut, thus making connection to the crankshaft.

Energizing the clutch coil creates lines of magnetic force from the poles of the electromagnet through the armature, drawing it toward the shoe plate or rotor that is a part of the pulley assembly. The solid mounting of the pulley prevents the pulley from moving in a lateral direction; however, the armature can move until it contacts the rotor. Magnetic force locks the rotor and the armature plate together. This solid connection then allows the pulley to rotate the compressor crankshaft and operate the compressor. Compressor operation will continue until the electrical circuit is broken to the clutch coil, when the magnetic force is de-energized. The rotor and armature then separate, and the pulley rotates freely without rotating the compressor crankshaft.

Slots are machined into both the armature and the rotor to concentrate the magnetic field and increase the attraction between the two when energized. Some scoring and wear is permissible between these plates. However, it is important that full voltage be available to the clutch coil as low voltage will prevent a full buildup of magnetic flux to the plates.

The correct spacing between the pulley and the coil on stationary coil models must be maintained to prevent the pulley from dragging against the coil. Correct spacing must also be maintained between the rotor and the armature.

Too close a clearance will allow the two plates to contact each other in the OFF position, while too wide a space can prevent the rotor from contacting the armature solidly in the ON position. Either of these variations can cause a serious clutch failure.

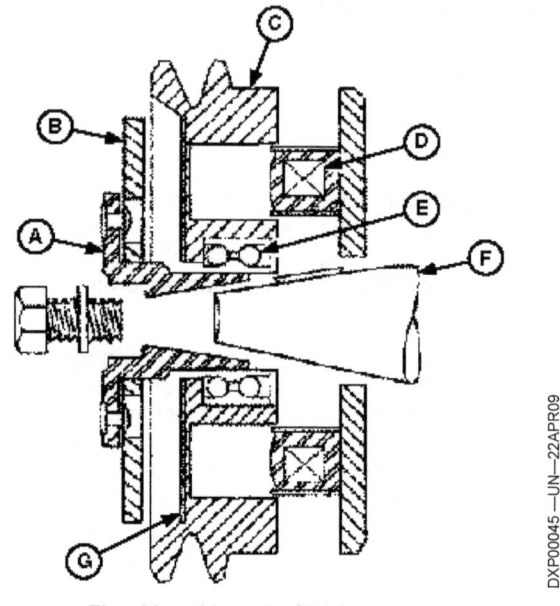

Fig. 23 — Magnetic Clutch

A—Hub
B—Armature
C—Pulley
D—Clutch Coil
E—Bearing
F—Compressor Crankshaft
G—Rotor Plate

Also be sure that the mating surfaces are not warped from overheating.

Always use the correct tools when performing any service operations, because damage to closely fitted parts may result from too much hammering or prying.

NOTE: For information on the electrical part of these clutches, refer to the FOS manual on "Electrical Systems."

Bypass Systems

Hot Gas Bypass

The control of evaporator pressure and temperature by metering a small amount of hot gas from the high side of the compressor (Fig. 24) has been used successfully many years in commercial refrigeration. General Motors engineers have also been very successful in adapting this same process to the control of their larger automotive air conditioning units.

Other manufacturers have attempted to adapt this type of control to their units, but soon dropped it in preference to some other type. Most have not gone beyond the experimental stage.

Solenoid Bypass

Some automotive products use a solenoid bypass to control the evaporator pressure and temperature. The thermal switch is attached to the suction pressure line at the evaporator outlet manifold. The electrical contacts in the switch are connected in series with the temperature control switch and the solenoid bypass valve winding. In the normal position, the contacts are closed.

A decrease in the temperature of the refrigerant gas leaving the evaporator will cause a thermal blade to bend and open the electrical circuit to the solenoid valve when the temperature reaches approximately 25°F

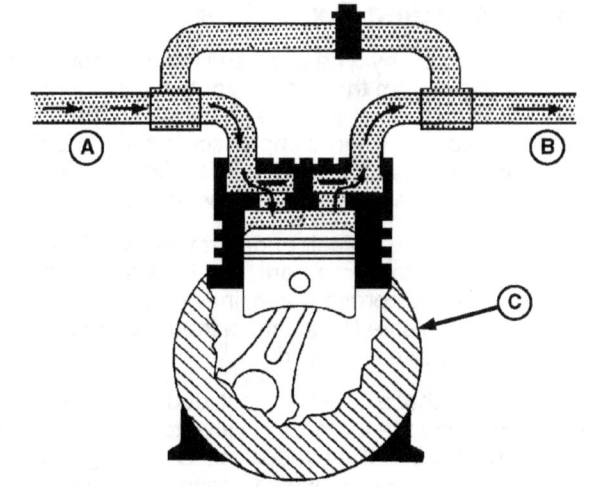

Fig. 24 — Hot Gas Bypass System

A—Low Side
B—High Side
C—Compressor

(–4°C). Opening the solenoid valve allows a charge of hot refrigerant gas to flow into the evaporator. When the temperature increases again to approximately 40°F (4°C), the contacts close and the solenoid energizes, closing the bypass valve.

Basic System: How It Works

Suction Throttling Regulators

Suction throttling is any type of control used to regulate the flow of refrigerant from the evaporator to the compressor. This control is located at some point between the tail pipe of the evaporator and compressor in the low side or suction line as it is commonly called (Fig. 25). These devices are used mainly on automotive systems.

Sometimes it may be difficult to determine exactly what type of control is used on a particular installation. If inspection of the compressor and lines does not show a control device, it is safe to assume a thermostat-controlled recycling clutch system is used.

We found earlier that a constant heat load and compressor speed would allow the expansion valve to meter an even flow of refrigerant into the evaporator. Changing any of these conditions, however, would change the refrigerant flow through the expansion valve. Without some way to regulate refrigerant flow, the evaporator cooling will become excessive and freeze the moisture, condensing on the coils. This results in evaporator icing or freeze-up and requires that the ice on the evaporator be melted before it can resume cooling.

Installation of a suction throttling control device will compensate for these varying conditions. As the pressure and temperature within the evaporator drop, spring pressure forces a valve within the regulating device toward its seat, retarding the flow of refrigerant. This action increases the pressure back into the evaporator, raising both pressure and temperature. The temperature

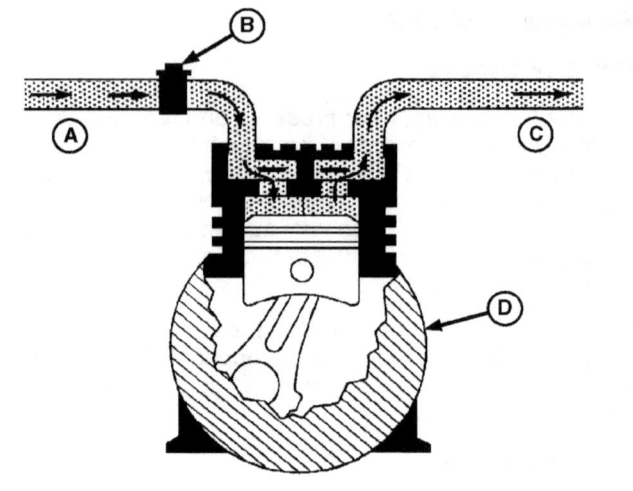

Fig. 25 — System Regulated by Suction Throttling

A—Low Side
B—Suction Throttling Regulator
C—High Side
D—Compressor

will not be allowed to drop low enough to freeze over the evaporator coils. A point of balance will soon be reached whereby the refrigerant flow past the valve in the regulator will reach a pressure and temperature sufficient to maintain cooling without evaporator freeze-up.

Continued on next page

Modulator Valve

The **modulator valve** (Fig. 26) limits and maintains a minimum pressure in the evaporator. It works like a hydraulic relief valve except that it has a suction relief.

The valve assembly is simple in operation. In Fig. 26, note that the valve has been divided into three functional areas.

The middle section is connected to the compressor suction valve. When the thermostatic expansion valve closes, the suction pressure will increase and compress the sealed bellows.

When the suction pressure becomes low enough, the bellows will force the valve disk open. This will allow the refrigerant to bypass the evaporator, thus stopping any further decrease in suction pressure.

When the thermostatic expansion valve at the evaporator inlet opens, the suction pressure will be relieved and the bypassing of refrigerant through the modulator valve will be stopped.

Located at the top of the modulator valve is a manual control plunger. This plunger is connected by a cable to the operator's temperature control. The manual control plunger regulates the range at which the valve opens and closes.

When the temperature control is pulled out for maximum cooling, the valve is fully closed and will require a low suction pressure for the refrigerant to bypass.

When the control is pushed in, the valve will be manually forced open, causing refrigerant to bypass freely through the modulator, destroying the suction pressure in the evaporator.

Other Types of Suction Throttling Regulators

Some other types of suction throttling regulators are:

- Robotrol Valves
- Evaporator Pressure Regulator (EPR)
- Suction Throttling Regulator (STR)

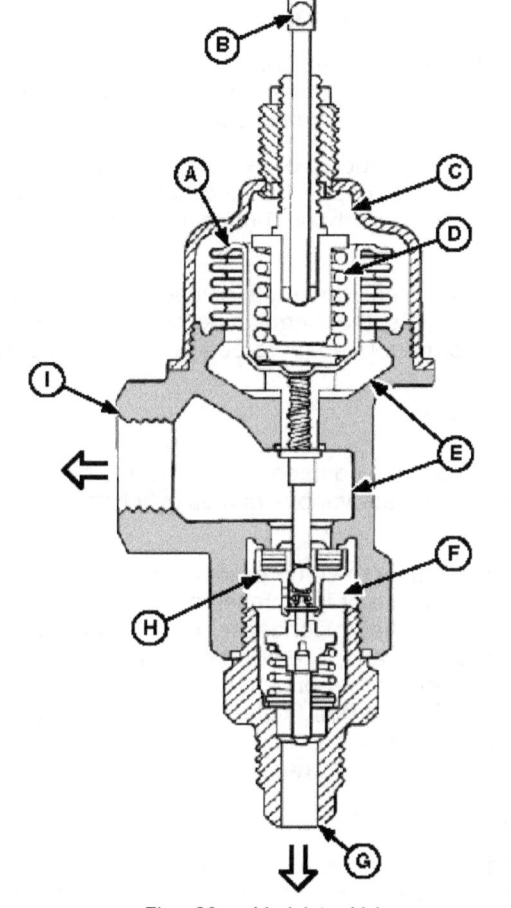

Fig. 26 — Modulator Valve

A—Sealed Bellows
B—Manual Control Plunger
C—Upper Section
D—Spring
E—Middle Section
F—Lower Section
G—Inlet (from Inlet Side of Expansion Valve)
H—Valve Disk
I—Outlet (to Compressor Inlet)

These regulators are used primarily in automotive systems.

Lines and Connections

Refrigerant lines carry refrigerant between the major components of the refrigeration system. They are joined to the components by connections.

Lines may be constructed of reinforced synthetic rubber, steel, aluminum, or copper. Connections (Fig. 27) are made with synthetic rubber O-rings, flare fittings, or hose clamps.

Some fittings are equipped with self-sealing couplings or "quick disconnects" that permit a new component such as an evaporator or condenser to be shipped with a partial charge.

Location

Listed below are the components connected by the major lines, followed by the various names by which these lines are known.

- Evaporator outlet to compressor inlet: **Suction line** (low-pressure line or low-pressure vapor line).
- Compressor outlet to condenser inlet: **Discharge line** (high-pressure vapor line or pressure line).
- Receiver-dehydrator outlet to thermostatic expansion valve inlet: **Liquid line** (high-pressure liquid line).
- Compressor outlet to evaporator outlet: **Hot gas bypass line** (hot gas line).

Operation

When the system is operating properly, the lines should be at the following general temperatures to the touch:

- Suction Line — Cool
- Discharge Line — Hot
- Liquid Line — Warm
- Hot Gas Bypass Line — Warm to hot (when bypassing refrigerant)

Diagnosis

Restrictions or kinks in the refrigerant lines may be indicated as follows:

Suction line — low suction pressure at the compressor, low discharge pressure, little or no cooling.

Discharge line — compressor relief valve opens.

Liquid line — low discharge pressure, low suction pressure, no cooling.

Hot gas bypass line — low suction pressure, possible evaporator icing.

Service

Plugged screens or hoses that leak or are damaged should be replaced. Hoses and lines should always be protected from rubbing against sharp metal surfaces, moving parts, or hot engine parts.

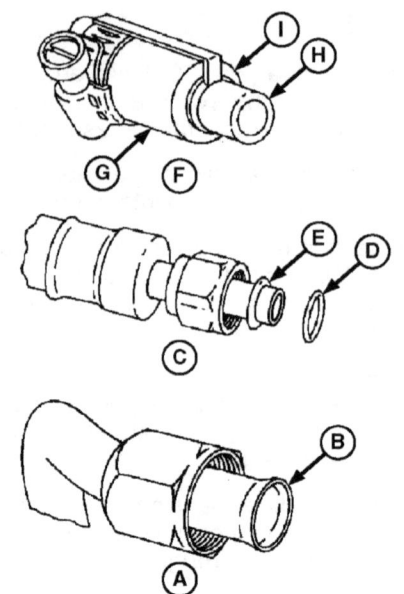

Fig. 27 — Types of Line Connections

A—Flare Fitting
B—Seal Seat
C—O-Ring Fitting
D—O-Ring
E—Upset Flange
F—Hose Clamp Fitting
G—Hose
H—Pipe
I— Bead on Pipe

The proper oiling and tightening of connections is very important. Always tighten the line connections to the torque recommended in the Technical Manual.

IMPORTANT: Use only approved components for the system being repair or replaced. Failure to do so can cause serious and costly damage to the air conditioning system.

It is important that hoses, O-rings, and seals be correct for the system being used. R-12 systems use nitrile rubber (NBR) compressor seals and hose connector O-rings. Hydrogenated nitrile rubber (HNBR) O-rings should be used on hose connections on R-134a systems. The choice of materials for compressor seals is made by the compressor manufacturer.

R-134a systems require hose material to be compatible with this refrigerant and lubricant. Since the system is likely to operate at higher pressures and the refrigerant tends to diffuse more easily than R-12 does through most hose materials, reinforced nylon core tube with butyl rubber cover hose is recommended.

Circulating the refrigerant in the system periodically throughout the year will help to lubricate seals and gaskets to ensure a sealed system. In winter, do this by turning over the compressor pulley by hand a few times.

Basic System: How It Works

Test Yourself

Questions and Problems

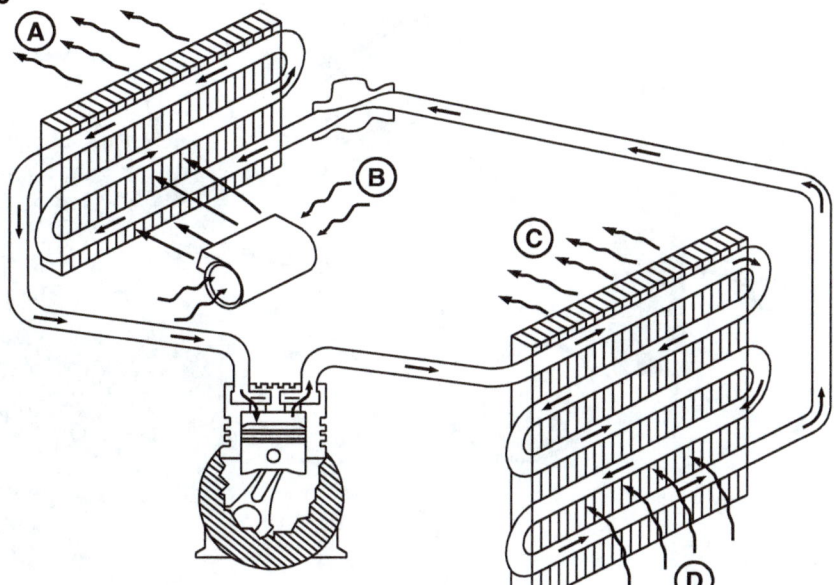

Fig. 28 — Diagram of Basic Air Conditioning System (See Question 1)

A—Cold Air
B—Warm Air
C—Hot Air
D—Air In
E—High-Pressure Liquid
F—High-Pressure Gas
G—Low-Pressure Liquid
H—Low-Pressure Gas

NOTE: An animated display of how the components in a basic Air Conditioning System works is available on the Instructor Art CD.

1. Mark up Fig. 28, "Diagram of Basic Air Conditioning System":

   a. Label the four basic parts of the system.

   b. Draw a line through the diagram at the correct place and label the "high side" and "low side" of the system.

   c. Use red and blue pencils and color the refrigerant passages as coded on the diagram to show high or low pressure and gas or liquid.

2. The compressor actually has two jobs. The first is compressing the refrigerant gas. What is the second?

3. What does compressing the refrigerant gas do to its heat content?

4. (Fill in the blanks.) Air passing over the condenser coils carries off _____ and changes the refrigerant from _____ to _____ .

5. Compare "ram air" to "forced air."

6. Which side of the expansion valve will be "cold" — inlet or outlet?

7. (Fill in the blanks.) Air passing over the evaporator coils changes the refrigerant _____ to _____ . As a result, heat is _____ .

8. To get the coldest air temperature from the evaporator, should the blower fan be normally run at its lowest or highest speed?

9. What is the purpose of the desiccant sealed inside the receiver-dryer?

See "Answers to Chapter 3 Questions" on page B-1.

Service Equipment

Introduction

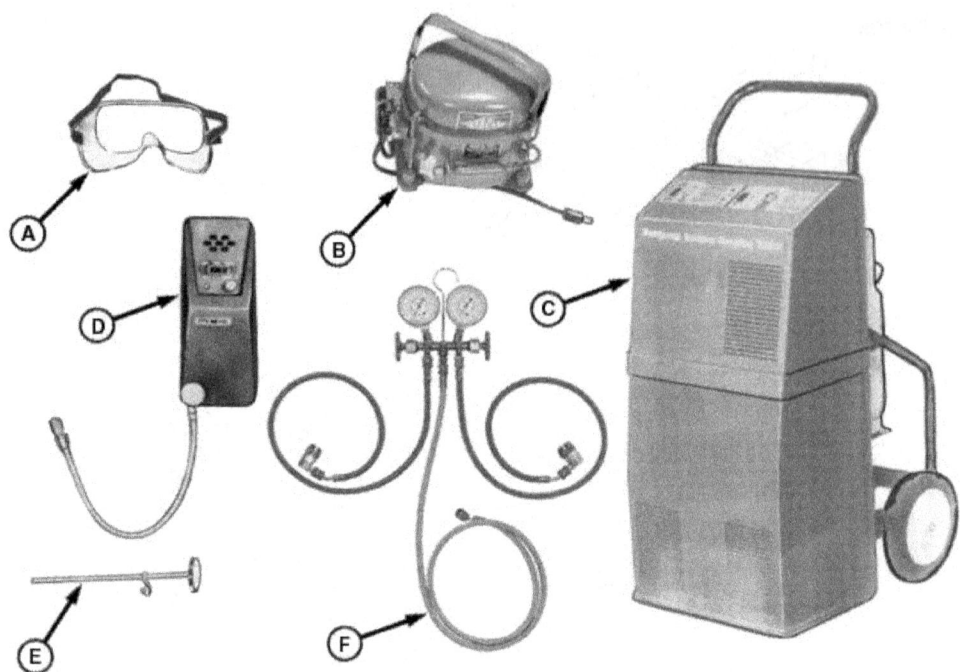

Fig. 1 — Complete Air Conditioning Service Kit

A—Plastic Goggles
B—Vacuum Pump
C—Refrigerant Recovery and Recycling Station
D—Leak Detector
E—Thermometer
F—Gauge and Manifold Set

The tools required to test and service an air conditioning system are shown in (Fig. 1).

Here we will describe only the operation of the various equipment. In Chapters 5 through 8 we will show the actual use of the tools in testing and servicing the system.

Refrigerant Recovery and Recycling Station

A **refrigerant recovery and recycling station** (Fig. 2) must be used to recover refrigerant from an air conditioning system to prevent the escape of refrigerant to the atmosphere. Release of refrigerants to the atmosphere deteriorates the earth's protective ozone layer.

There are also refrigerant recovery, recycling, and recharging stations available in one unit, which are equipped with high- and low-pressure gauges, vacuum pump, and recharging capabilities. When the refrigerant recovery, recycling, and recharging station is connected to the air conditioning system, you can diagnose the system, evacuate and then recycle the refrigerant. After repairs are made, the refrigerant recovery, recycling, and recharging station can purge, evacuate, and recharge the system.

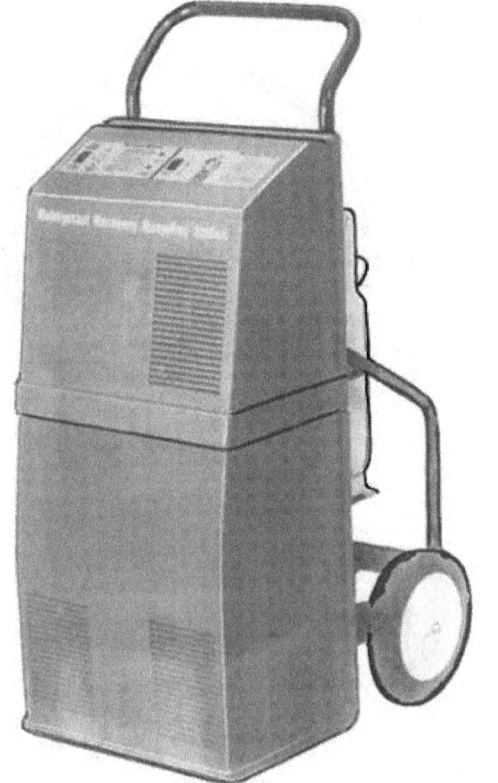

Fig. 2 — Refrigerant Recovery and Recycling Station

Service Equipment

Gauge and Manifold Set

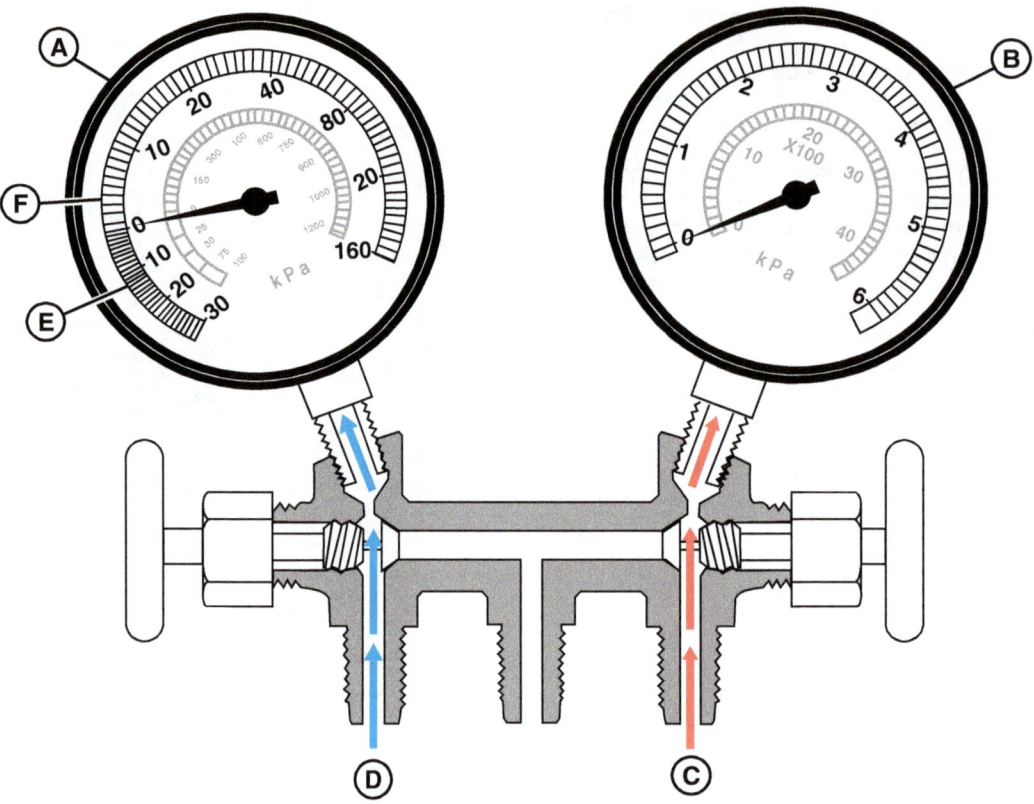

Fig. 3 — Refrigerant Flow to Gauges in Performance Test — Both Hand Valves Closed

A—Compound Gauge (Low Side)
B—High-Side Gauge
C—High-Side Service Connector
D—Low-Side Service Connector
E—Vacuum (Hg)
F—Pressure

Accurate testing requires the use of a test gauge set connected to the high and low sides of the air conditioning system. With these gauges, the service technician can accurately pinpoint trouble within the system as well as determine if the system is operating as it should.

The **gauge manifold set** (Fig. 3) is composed of a low-side or compound gauge, a high-side gauge, and the manifold to which the gauges are connected with a hand valve at each end.

Fittings on gauge sets for R-134a systems have male threaded ends, while R-12 system gauge sets have female connections.

Following is a brief description of the gauges and their requirements:

Compound Gauge (Low Side)

The **compound gauge** (Fig. 3) derives its name from its function. It can register both pressure and vacuum. All air conditioning systems can, under certain conditions, drop from a pressure into a vacuum on the low side. It is necessary that a gauge be used that will show either pressure (psi and kPa) or inches of mercury vacuum (Hg).

The vacuum side of the gauge must be calibrated to show 0–30 inches (0–762 mm) Hg. The pressure side of the gauge must be calibrated to register accurately from 0 pressure to a minimum of 60 psi (414 kPa). The maximum reading of the pressure should not exceed 160 psi (1103 kPa). Practically all readings of the low side of the system will be less than 60 psi (414 kPa) with the system in operation.

Each service technician can choose the scale reading preferred. To accurately convert pressures to temperatures in the system, the gauge should be calibrated to a low enough scale that it will not be difficult to obtain an accurate reading. The higher the pressure scale, the more difficult it becomes to get an accurate pressure-temperature conversion.

Continued on next page

Service Equipment

High Pressure Gauge (High Side)

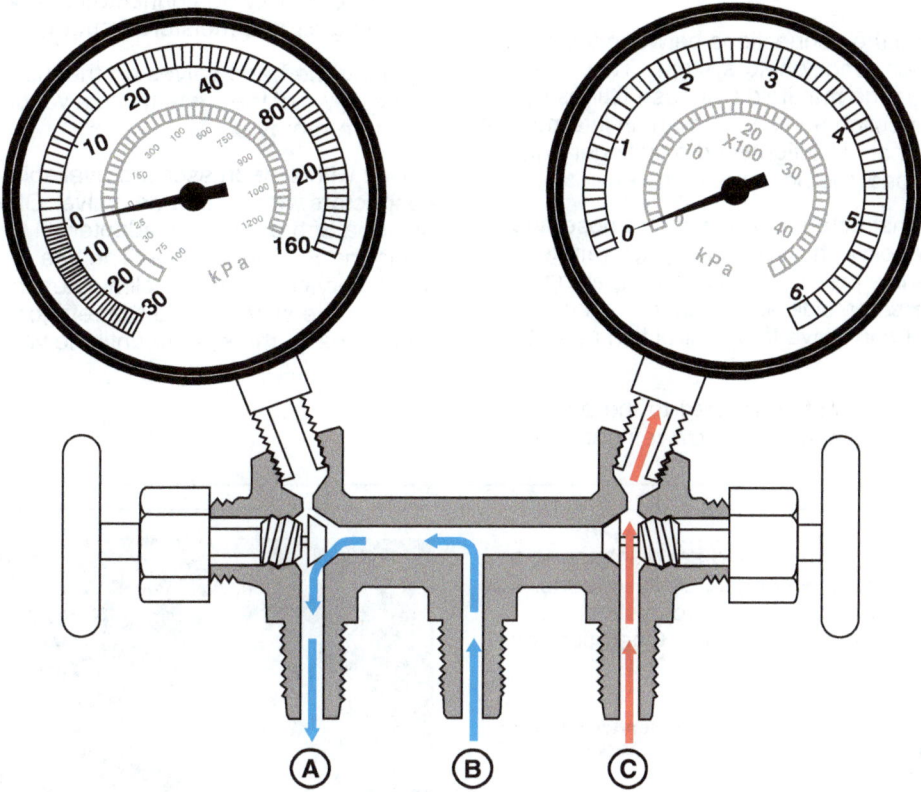

Fig. 4 — Refrigerant Flow to Gauges and through Manifold while Adding Refrigerant

A—From Low-Side Service Connector
B—Refrigerant Source
C—From High-Side Service Connector

The high pressure gauge is used to determine pressures in the high side of the system. The gauge is calibrated to register accurately from 0 pressure to a minimum of 300 psi (2070 kPa). A few systems operate under high head pressure during normal operation conditions. This is why the high pressure gauge should have a reading of at least 600 psi (4140 kPa).

Gauge Manifold

The **gauge manifold** mounts the high- and low-side gauges and connects the gauges into the high and low sides of the system by means of test hoses. The gauges connect to the upper part of the manifold through holes drilled and tapped to a 1/4-inch pipe thread. Test hose connectors below the gauges on the lower side of the manifold direct the refrigerant through the manifold to the gauges to obtain pressure readings.

A center test hose connector on the lower side of the manifold is connected to both pressure gauges and the test hoses by a passage in the manifold. Refrigerant flow into the high and low sides is controlled by a shutoff hand valve at each end of the manifold.

With both hand valves in the closed position (Fig. 3), refrigerant will be shut off from the center test hose fitting but will flow to the gauges.

Opening the low-side gauge will open the low-side refrigerant to the center test hose connection and the low-side gauge (Fig. 4).

Opening the high-side hand valve will allow refrigerant to flow through the passage and out the center test hose connector and at the same time continue to the high gauge to register pressure reading.

By opening and closing the hand valves on the manifold, the following procedures can be performed:

- Recovering excess refrigerant from system
- Bleeding air from the hoses
- Recovering refrigerant before service work
- Removing air and moisture during pump-down
- Filling system with refrigerant

IMPORTANT: Do not release refrigerants to the atmosphere. Use a suitable recovery system such as the one shown in Fig. 2.

All of these procedures will be explained in Chapters 7 and 8.

Continued on next page

Service Equipment

Test Hoses

The **test hoses** are the connections between the gauge manifold and the air conditioning system. They are connected to the gauge manifold test hose fittings by use of a screw-on connection and sealed with an internal seal. Hose connectors should be tightened only finger-tight as this is sufficient to seal the hose onto the seal.

The manifold is constructed so that the test hose and connector directly below the gauge will pass refrigerant to that gauge to show pressure readings. Opening the hand valve on the same side as the gauge is the only way the refrigerant can move in any direction other than to the gauge.

The center test hose is not connected to the air conditioning system. It is used to allow refrigerant to purge from the system using a recovery and recycling station, or it may be connected to a vacuum pump for removing air and moisture from the system.

Opening the hand valves on the manifold will control pump-down of the system into a vacuum for more effective moisture removal.

Hoses with a depressor are available to fit the service connectors with a Schrader valve. Other hoses require the use of a Schrader valve adapter on the connectors before using the Schrader valve. The use of the Schrader valve in the service connector eliminates the need for a service valve in the system, and the refrigerant is effectively sealed inside the system until the valve is opened.

Fig. 5 shows a gauge and manifold set discharging a system through the compressor service valve ports. For instructions on installing gauge set into system, see "Installing Gauge Set to Check System Operation" on page 07-1.

A—Low-Side Hose
B—High-Side Hose
C—High-Side Valve
D—Center Service Hose
E—Low-Side Valve

Fig. 5 — Typical Gauge and Manifold Set

Service Valves

The compressor is equipped with **service valves**, which are used as an aid in servicing the air conditioning system. The manifold gauge set is connected into the system at the service valve ports, and all procedures such as evacuating and charging the system are carried on here through the gauge and manifold set.

Most compressors are equipped with two service valves. One services the high side, while the other services the low side. The high-side service valve is quickly identified by the smaller discharge hose routed to the condenser, while on the low-side valve the larger hose comes from the evaporator.

Since all valves are the same, we will be concerned here with the operation of one valve in the system.

The valves described here are the hand shutoff type. Many air conditioning systems now use only one valve having a shutoff feature or one valve having no shutoff feature. The gauge hoses are still connected to the

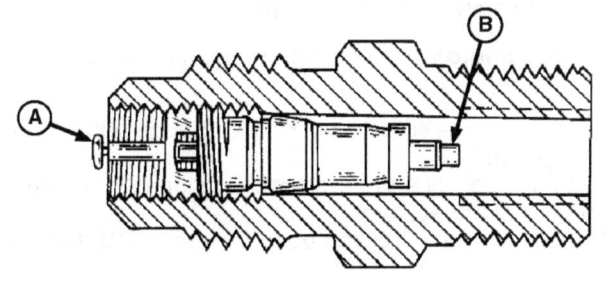

Fig. 6 — Schrader Valve

A—Pin B—Valve

service valve fitting, in which a Schrader valve (Fig. 6) is incorporated. When the fitting in the end of the service hose is screwed onto the Schrader valve, a pin is depressed in the center of the valve, allowing pressure to be read on the gauges. When the fitting is removed, the valve closes to hold refrigerant in the system.

Continued on next page

Service Valves — R-12 System

The **hand shutoff type of valve** (Fig. 7) has three positions:

- Front-seated—All flow shut off (A). Gauge port out of the system.
- Back-seated—Normal system operation (B). Gauge port out of the system.
- Mid-positioned—Cracked open for testing (C). Gauge port in the system.

We will discuss each position and determine at what points refrigerant will be allowed to flow.

Shut Off Refrigerant Flow

In this position we refer to the service valve as being in the front-seated position (A).

You can see that the refrigerant is trapped in the hose end of the service valve. The gauge port fitting is toward the atmosphere. By following the path through the valve, you can see that the gauge port only connects to the compressor. If the compressor were run with the service valve in this position and the gauge port capped, serious damage would occur to the compressor. There would be no area to pump into.

IMPORTANT: Never operate a compressor with the service valves in closed or front-seated position. This will damage the compressor.

Normal Refrigerant Operation

The service valve is in the back-seated position (B). The compressor and hose outlet are connected, and refrigerant is free to flow if the compressor is started. Now the gauge port is closed off and pressure readings cannot be taken. All service valves should be in this position when the system is operating normally.

Testing Operation

The service valve is in the cracked, or mid-position (C). Now the system can be operated and the pressures recorded through the gauge port openings.

Remember, however, that the valves must always be back-seated again before attempting to remove the gauge hose from the service valves. Failing to do so will result in loss of refrigerant. The mid-position (C) shows the presence of refrigerant at all outlets of the service valve for testing.

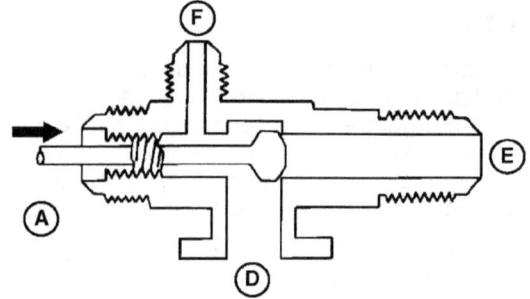
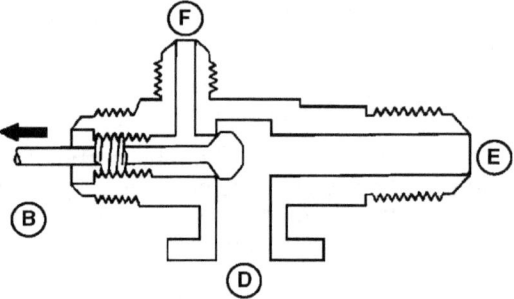
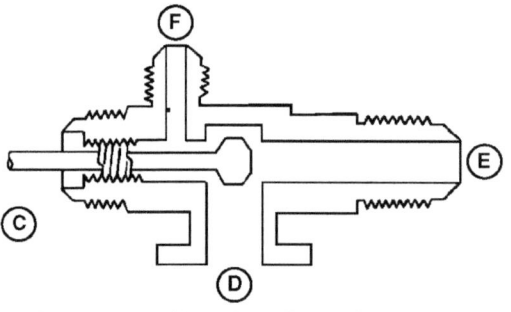

Fig. 7 — Service Valve (Hand Shutoff Type for R-12 System)

A—Front-Seated—All Flow Shut Off
B—Back-Seated—Normal System Operation
C—Mid-Positioned—Cracked Open for Testing
D—To Compressor
E—To Hose
F—To Gauge Port

Continued on next page

Service Equipment

The location of service valves at the compressor for a typical system is shown in Fig. 8.

Valve position is controlled by rotating the valve stem with a service valve wrench.

A—Service Valve Caps
B—Compressor
C—Gauge Port Caps
D—Suction Service Valve (Low Side)
E—Discharge Valve (High Side)

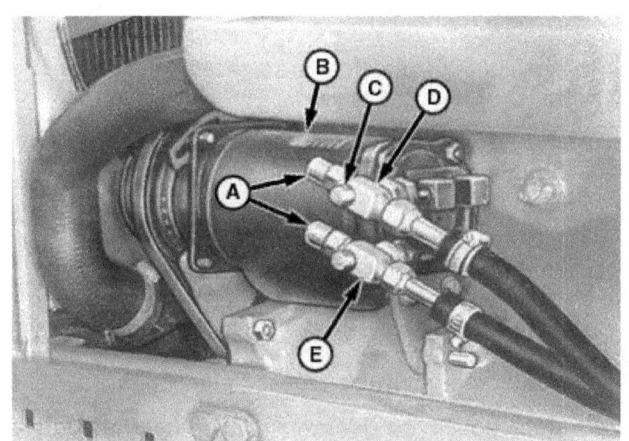

Fig. 8 — Location of Service Valves in a Typical System

Service Valves—R-134a System

New and unique service hose fittings have been specified for R-134a systems. Their purpose is to avoid cross-mixing of refrigerants and lubricants with R-12-based systems. The service ports on the system are quick disconnect type with no external threads. They do contain a Schrader-type valve as shown in Fig. 9. The low-side fitting has a smaller diameter than the high-side attachment.

A—System Service Port
B—Quick Connect Hose End Fitting with Integral Shutoff Valve
C—Service Hose Connection
D—Depressor Pin

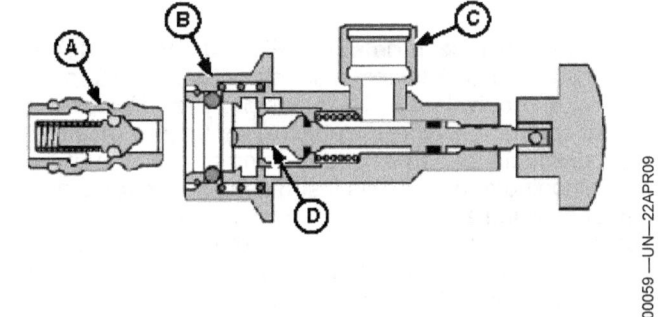

Fig. 9 — Service Valve — R-134a System

Leak Detectors

Several types of leak detectors are available to the service technician:

- Colored Dye Additive
- Liquid Detergent-Type Detector
- Electronic Leak Detector

A **colored dye additive** is available, which is added in the refrigerant. Operation of the system will show coloration at the point of leakage. A very slight leak requiring several weeks or even months to bleed off enough refrigerant to affect system cooling can often be located using this additive when other methods of leak detection fail.

A **liquid detergent-type detector** may be used around connections and any external point that might be a source of leak for the R-12. Escaping refrigerant will cause the liquid to bubble, indicating a leak. Any parts that are not accessible, such as the coils in the condenser and the evaporator, cannot readily be coated with this liquid to check for leaks.

The **electronic leak detector** (Fig. 10 and Fig. 11) is a sensitive leak detector. Most electronic detectors can detect an equivalent of 1/2 ounce per year. However, the initial cost of this type of detector has been a deterrent to individuals and small shops doing a minimum of air conditioning service. This instrument is electronic and must be handled with care to give accurate results. When cared for properly, the electronic detector will quickly and accurately locate leaks that are almost impossible to locate with other types of detectors.

Leak detection must be performed with the system under pressure to obtain accurate results. Very small leaks often require that the system pressure be increased above normal before they can be located. A 50% charge of refrigerant in the system is enough to locate most leaks. Occasionally, a stubborn small leak will require overcharging the system.

The **high side** of the system might require leak testing while in operation with air flow restricted to the condenser to raise the high-side pressure above normal.

The **low side** is checked in the OFF position with the pressures equalized in both sides of the system.

When you use leak detectors, don't move the sniffer or snorkel faster than 1 inch (25 mm) per second.

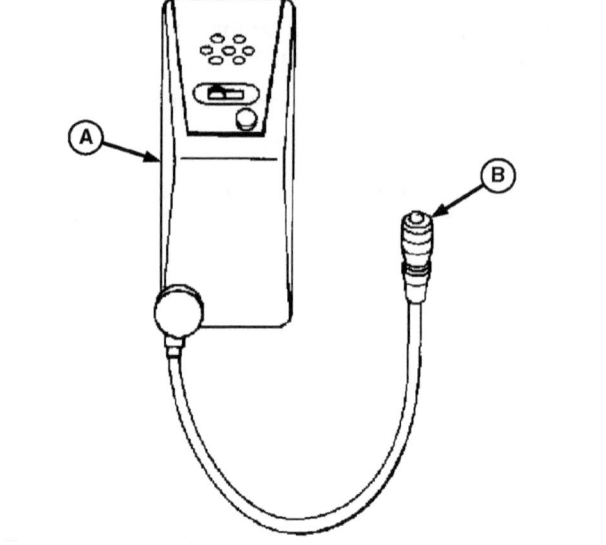

Fig. 10 — Electronic Leak Detector

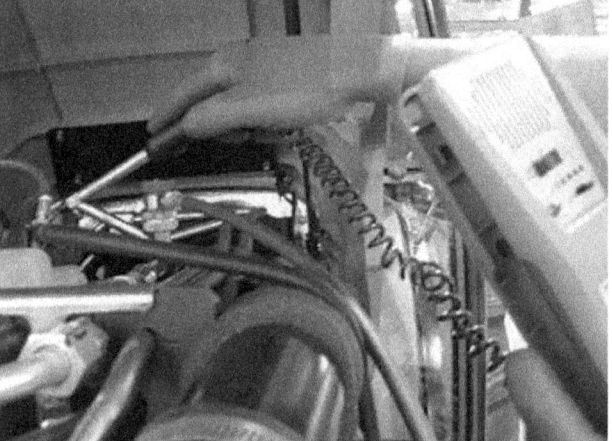

Fig. 11 — Leak Testing System Using Electronic Leak Detector

A—Detector Unit B—Sampling End

Many leaks can be found by looking for small oily spots or film at the source of the leak.

Vacuum Pump

The **vacuum pump** is used to evacuate air from the system and is located in the **Refrigerant Recovery and Recycling Station** (Fig. 12).

When the system is depressurized and opened for service, air enters the openings before they can be capped. To remove this air (and its harmful moisture), the system must be evacuated. This is done by removing air until a vacuum is created. Detailed procedures are given in Chapter 8.

A—From Evaporator
B—Compressor
C—To Condenser
D—Gauge Set
E—High
F—Low
G—Refrigerant Recovery and Recycling Station

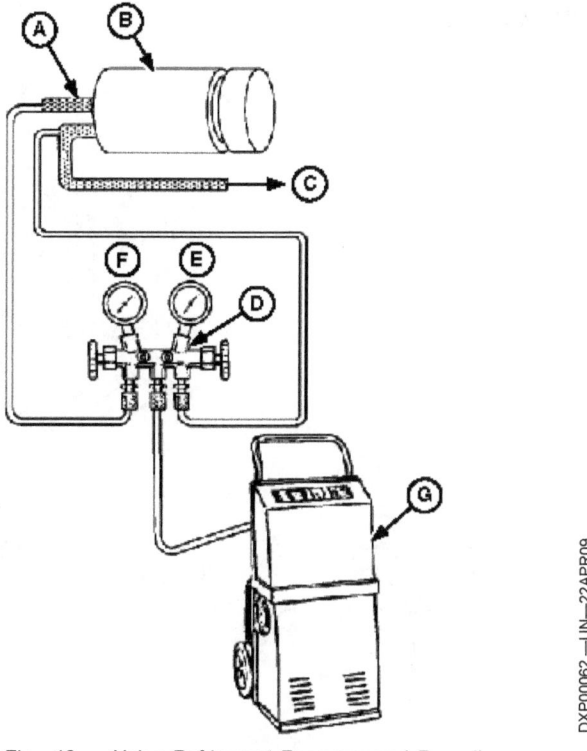

Fig. 12 — Using Refrigerant Recovery and Recycling Station to Evacuate the System

Other Service Tools

Since the air conditioning system is activated by electrical controls, some electrical test equipment such as a voltmeter will also be needed to check for faulty wiring and other problems.

Test Yourself

Questions

1. What three basic tools are required to test and service an air conditioning system?

2. Why are two gauges needed to test the pressure in the system?

3. On what component are the system service valves located?

4. What type of leak detector is most sensitive to small leaks?

5. What happens when refrigerant is passed over an open flame?

6. What equipment is used to remove air from the system?

See "Answers to Chapter 4 Questions" on page B-1.

Inspecting the System

Introduction

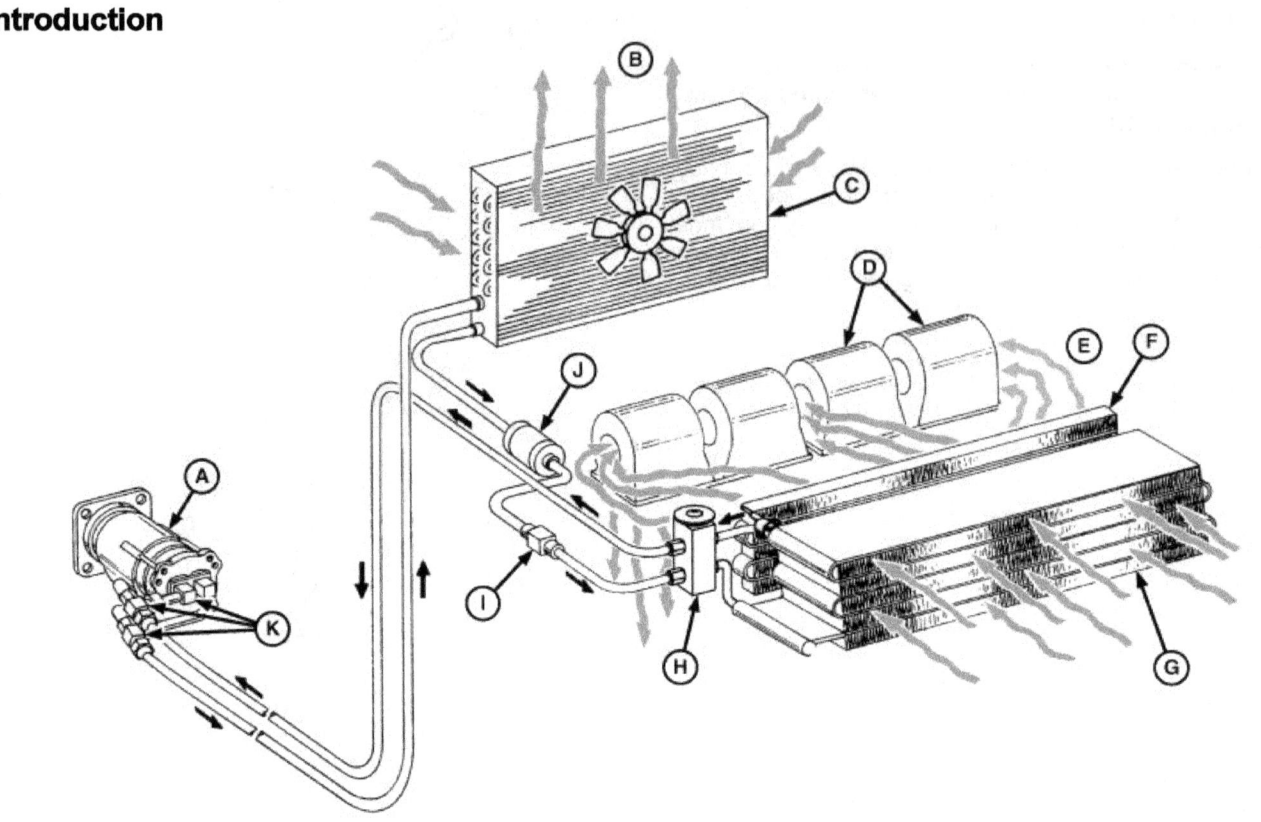

Fig. 1 — Air Conditioning System for Typical Tractor with Cab

A—Compressor
B—Air Out
C—Condenser
D—Motors and Blowers
E—Air In
F—Heater Core
G—Evaporator
H—Expansion Valve
I—Sight Glass
J—Receiver-Dryer
K—Compressor Manifold and Service Valves

A seasonal check of the air conditioning system (Fig. 1) is very important in revealing troubles early before they cause a failure.

A **performance test** is the only certain way the complete system can be checked for efficient operation (see "Installing Gauge Set to Check System Operation" on page 07-1). Whenever possible, the system should be given this test before work is begun on the system.

Many times, however, the system is completely inoperative, and repairs must be performed before it can be properly tested. The test can uncover further work that must be performed before the system is brought to full operating efficiency.

The performance test should always be performed after repair work has been done and before the machine is released to the customer. The service technician performing this test carefully will ensure that the repairs have been properly performed and that the system will operate satisfactorily.

A good performance test includes a thorough examination of the outside of the system as well as the inside. Many related parts are overlooked because it is felt they are of no importance to the system actually cooling the inside of the cab. But often these outside parts have a direct bearing on the operating efficiency of the unit.

For this reason, a thorough **visual inspection** of the complete system should be performed, followed by an **operating inspection** of the system.

Inspecting the System

Visual Inspection of the System

Visually Inspect the Following:

1. **Compressor drive belts tight, not worn or frayed, and aligned with pulleys.** The compressor belt (Fig. 2) is subjected to a heavy load during operation. This is especially true when the head pressures build up in excess of 200 psi (1380 kPa) in hot weather operation. The belt must be in excellent condition to withstand the strain of heavy loads. If the pulleys are not properly aligned, extreme side wear to the belt and pulleys will result. Too tight a belt tension will result in strain to the bearings of units operated by the compressor belt. Too loose a belt tension will result in bolt slippage and poor performance. A belt tension gauge eliminates guesswork in tightening the compressor belt. If a belt tension gauge is not available, tighten until there is a 3/8- to 1/2-inch (10 to 13 mm) deflection between any two pulleys that are farthest apart.

2. **Compressor brackets and braces tight and not cracked or broken.** Mounting bolts can work loose and bracket end braces often break under the vibrations and strain of operation. Failure to inspect and repair any damage at these points can result in early system failure.

3. **Hoses or copper lines not chafing or leaking.** Grommets and rubber pads that were originally installed to protect the hoses from contact with metal parts may deteriorate or loosen. Exposing the hose or line to constant rubbing and chafing can cause deterioration and allow the refrigerant to escape. To prevent damage, install some type of protective material.

4. **Condenser clean and properly mounted.** Insects and dirt clog the condenser and radiator and stop air movement. Any blocking of full air flow over the condenser and radiator coils must be corrected to allow proper condensing action of the system.

5. **Evaporator is clean.** The evaporator condenses moisture, which in turn traps dust and lint on the side where the air enters. The blower or fan can be effective only when evaporator passages are clear. Dust and lint should be removed.

6. **Compressor oil level correct.** Most compressors do not have a provision to check the oil level without disconnecting the compressor from the system. Make an oil level check only with the system discharged. Do not overfill the system with oil, because flooding

Fig. 2 — Typical Compressor Drive Belt and Pulley

A—Pulley
B—High-Side Port
C—Low-Side Port
D—Compressor
E—Compressor Drive Belt

of the condenser and evaporator will result. See "Refrigeration Oil" on page 02-5 and "Checking and Adding Oil to Axial Piston Compressors" on page 07-7.

7. **Air ducts and louvers operating smoothly.** Operate all mechanisms to check for free operation without binding and sticking.

8. **Blower motor operating satisfactorily.** Operate blower motor at all speeds. If motor is noisy or fails at some speeds, repair it.

9. **Air filters are clean.** Many systems use a fresh air and a recirculating filter to clean the air before it goes to the evaporator coil. The filters must be removed and cleaned, because a clogged filter will seriously affect evaporator air flow.

10. **No visible leaks.** A small, oily spot usually indicates a refrigerant leak, as oil is carried out with the escaping refrigerant.

11. **Leak test the system.** A leak test will tell whether an oily spot indicates a leak. See "Leak Testing System Using Electronic Leak Detector" on page 07-13. This test can only be performed on systems that are operative. A unit that has lost its refrigerant must be partially charged before this test can be performed.

Operating Inspection of the System

An operating inspection of the system can be made for three factors:

- System fully charged
- Relation of temperatures at high and low sides of system are okay
- Evaporator outlet blowing cool air

Before making these inspections, operate the system for about 5 minutes to allow refrigerant in system to stabilize.

Checking System for Full Charge

1. Use the test gauges and the sight glass (if equipped) for this test. See "Installing Gauge Set to Check System Operation" on page 07-1 for installing gauges.

2. **High-side** or head pressure will normally read 150–270 psi (1035–1860 kPa) (10.3–18.6 bar) depending upon ambient air temperatures and the type of unit tested. (See Pressure-Temperature Table in this chapter.)

3. The sight glass (if equipped) should be free of bubbles after system has been operating for a few minutes.

4. **Low-side pressure** should read 1–30 psi (7–207 kPa) (0.1–2.1 bar) again depending on the air temperatures and the unit tested.

5. It is impossible to give a definite reading for all types of systems, as the type of control and component installation will influence the pressure readings of the high and low sides.

6. The high-side pressure will definitely be affected by the ambient or outside air temperature. A system that is operating normally will indicate a high-side gauge reading of 150–170 psi (1035–1172 kPa) (10.3–11.7 bar) with a 76–80°F (24–27°C) ambient temperature. The same system will register 210–230 psi (1450–1585 kPa) (14.5–15.9 bar) with an ambient temperature of 100°F (38°C). No two systems will register exactly the same, so allow for variations in head pressures.

Checking Relative Temperature at High and Low Sides of System

The **high side** of the system should vary from **hot** at compressor discharge valve to **warm** at expansion valve. A difference in temperature will indicate a partial blockage of liquid or gas at this point.

Fig. 3 — Air Conditioning System for Farm Machine with Cab

A—Evaporator B—Condenser

Pressure-Temperature Table	
Ambient Temp.	Normal High-Side Pressure
80°F (27°C)	150–170 psi (1034–1172 kPa)
90°F (32°C)	175–195 psi (1207–1344 kPa)
95°F (35°C)	185–205 psi (1235–1413 kPa)
100°F (38°C)	210–230 psi (1448–1586 kPa)
105°F (41°C)	230–250 psi (1586–1724 kPa)
110°F (43°C)	250–270 psi (1724–1862 kPa)

The **low side** of the system should be uniformly **cool** to the touch with no excessive sweating of the suction line or low-side service valve. Excessive sweating or frosting of the low-side service valve usually indicates the expansion valve is allowing an excessive amount of refrigerant into the evaporator. This could be caused by a loose terminal bulb on the outlet of the evaporator.

Checking Evaporator Output

If all the previous inspections have been performed carefully and components have been found to operate properly, a rapid cooling of the cab interior should result.

Use a thermometer to read ambient air temperature. After the air conditioning system has been running 10 to 20 minutes, obtain a temperature reading from the blower air duct outlet. The temperature difference should be as follows:

Ambient Air	Temperature Difference (min.)
Below 75°F (24°C)	20°F (11°C)
75–90°F (24–32°C)	25°F (14°C)
Above 90°F (32°C)	30°F (17°C)

Inspecting the System

Test Yourself

Questions

1. What is the normal range for low-side pressure in the system?

2. Fill in the normal high-side pressure ranges in the chart below.

Ambient Temp.	Normal High-Side Pressure
80°F (27°C)	
90°F (32°C)	
95°F (35°C)	
100°F (38°C)	
105°F (41°C)	
110°F (43°C)	

3. Match the temperatures below with the components listed.

System Component	Normal Temperature
1. Expansion valve	a. Hot to the touch
2. Low-side suction line	b. Warm to the touch
3. Compressor discharge valve	c. Cool to the touch

See *"Answers to Chapter 5 Questions"* on page B-1.

Diagnosing the System

Introduction

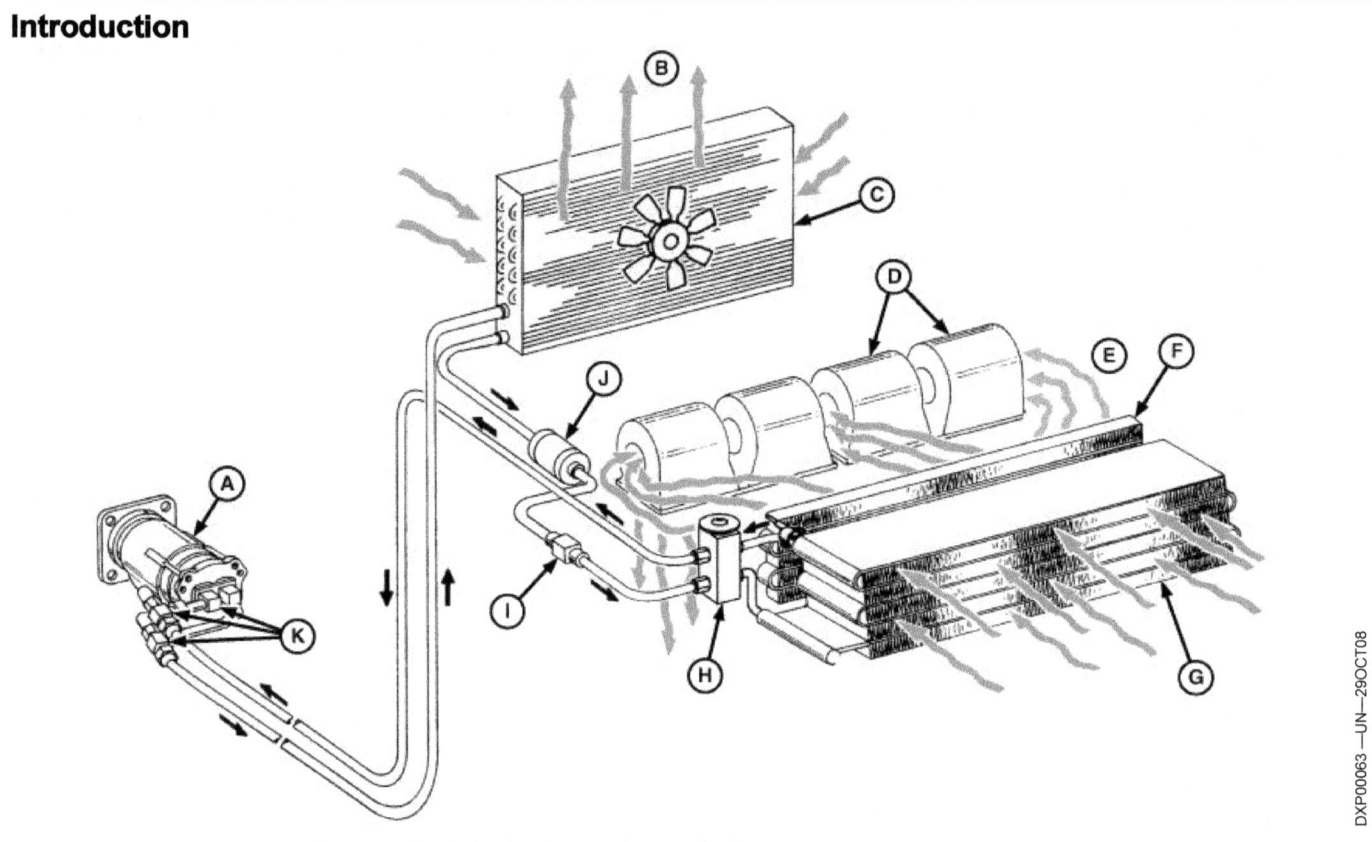

Fig. 1 — Basic Air Conditioning System for Farm or Industrial Tractor with Cab

A—Compressor
B—Air Out
C—Condenser
D—Motors and Blowers
E—Air In
F—Heater Core
G—Evaporator
H—Expansion Valve
I—Sight Glass
J—Receiver-Dryer
K—Compressor Manifold and Service Valves

Chapter 6 is divided into three sections.

- Troubleshooting Customer Complaints
- Flow Charts for Diagnosing the System
- Diagnostic Chart

Troubleshooting Customer Complaints

The following charts are offered as an aid in troubleshooting all kinds of air conditioning systems. All complaints in this section fall into three main categories:

- Electrical
- Mechanical
- Refrigeration

An inspection will tell which of these categories the problem falls into.

In many cases, a problem that causes an air conditioning system to malfunction requires little time to check out and repair. These possible causes should be the first to be examined and corrected.

For example, a complaint will likely be that the air conditioner produces no cooling. Before installing pressure gauges, take time to check a few possible causes. Go to "Customer Complaints — Troubleshooting Chart" on page 06-3 and review the causes for System Produces No Cooling.

If electrical components are operating, the causes under **Electrical** can be eliminated. You may want to check for a burned-out or disconnected clutch coil and solenoid, excessively burned electrical switch contacts in the thermostat, or defective sensing element.

Next check for **Mechanical** problems that could be inspected and repaired without attaching the pressure gauges — lines 1 and 2. These can be examined and corrected, if necessary, with little difficulty. Lines 3, 4, and 5 require installing the gauges.

In this example, it would be appropriate to then inspect for problems related to Refrigeration in the next section of the chart. Lines 1 and 2 under **Cause** require the least amount of time to inspect and repair.

If the problem is not discovered after inspecting these possibilities, it will be necessary to install the pressure gauges. The remaining mechanical and refrigeration problems may then be examined using pressure readings as a guide.

By using this type of logic to troubleshoot air conditioning systems, many problems can be repaired in a minimum amount of time. First, eliminate the possible causes of problems that are easiest to check and repair. Then, if the problem is not discovered, attach the pressure gauges and check the other possible causes of the problem. If none of the symptoms listed are detected, use the flow charts for a systematic diagnosis of the entire system. See "Flow Charts for Diagnosing the System" on page 06-6. Remember, these troubleshooting charts are a methodical procedure to locate the source of a problem and give a systematic examination to determine what the problem is and how to repair it.

NOTE: All troubleshooting information applies to air conditioning systems with the receiver-dryer placed between the condenser and evaporator.

Customer Complaints — Troubleshooting Chart

Trouble	Cause	Indications	Remedy
I. System Produces No Cooling	**Electrical**		
	1. Blown fuse.	1. Electrical components will not operate.	1. Replace fuse.
	2. Broken or disconnected electrical wire.	2. Electrical components will not operate.	2. Check all terminals for loose connections; check wiring for hidden breaks.
	3. Broken or disconnected ground wire.	3. Electrical components will not operate.	3. Check ground wire to see if loose, broken, or disconnected.
	4. Clutch coil or solenoid burned out or disconnected.	4. Compressor clutch or solenoid inoperative.	4. Check current flow to clutch or solenoid — replace if inoperative.
	5. Electrical switch contacts in thermostat are burned excessively, or sensing element defective.	5. Compressor clutch inoperative (applies to units having thermostatically controlled recycling).	5. Replace thermostat.
	6. Blower motor disconnected or burned out.	6. Blower motor inoperative.	6. Check current flow to blower motor — repair or replace if inoperative.
	Mechanical		
	1. Loose or broken drive belt(s).	1. Visual inspection.	1. Replace drive belt(s) and/or tighten to specification(s).
	2. Compressor partially or completely seized.	2. Compressor belt slips on pulley or the compressor will not turn when clutch is engaged.	2. Remove compressor for service or replacement.
	3. Compressor reed valves inoperative.	3. Only slight variation of both gauge readings at any engine speed.	3. Service or replace compressor reed valves.
	4. Expansion valve in open position.	4. Head pressure normal or high, suction pressure high, evaporator flooding.	4. Inspect valve's thermal bulb for corrosion and tension to tail pipe test expansion valve.
	5. Expansion valve stuck shut.	5. Head pressure low, suction pressure low.	5. Test expansion valve.
	Refrigeration		
	1. Broken refrigerant line.	1. Complete loss of refrigerant.	1. Examine all lines for evidence of breakage by external stress or rubbing wear.
	2. Fusible plug blown from receiver-dryer (does not apply to all units).	2. Complete refrigerant loss.	2. Examine fusible plug — if blown, replace receiver-dryer.
	3. Leak in system.	3. No pressure on high- and low-side gauges (applies to any system having complete loss of refrigerant).	3. Evacuate system, apply static charge, leak test system, and repair leak as necessary.
	4. Compressor shaft seal leaking.	4. Clutch and front of compressor oily; system low or out of refrigerant.	4. Replace the compressor shaft seal.
	5. Clogged screen(s) in receiver-dryer or expansion valve; plugged hose or coil.	5. High-side gauge normal or may read high. Low-side gauge usually shows vacuum or very low pressure reading. Frosting usually occurs at point of blockage.	5. Repair as necessary.

Continued on next page

Diagnosing the System

II. System Will Not Produce Sufficient Cooling	Electrical	NOTE: After completing repairs of any of the previous causes, system **must** have receiver-dryer replaced, evacuated, and charged.	
	1. Blower motor sluggish in operation.	1. Small displacement of air from discharge duct; blower motor possibly noisy.	1. Remove blower motor for service or replacement.
	2. Compressor clutch slipping.	2. Visual inspection.	2. Remove clutch assembly for service or replacement.
	3. Obstructed blower discharge passage or damaged blower fan.	3. Blower operates at high speed but air displacement very small.	3. Examine fan and entire discharge passage for kinks, obstructions, or failure of passage to open during installation. Correct as necessary.
	4. Clogged air intake filter or recirculating filter.	4. Too little air displacement by blower.	4. Remove air filters and service or replace, whichever is necessary.
	5. Outside air vents open.	5. Too little cooling at high ambient temperature.	5. Close air vents (adjust controls if necessary).
		NOTE: Some owners must be instructed on importance of keeping air vents closed when air conditioning unit is in operation.	
	6. Too little air circulation over condenser coils; fins clogged with dirt or bugs.	6. Too little cooling at discharge outlet; excessively high pressure gauge reading; engine temperature usually excessive.	6. Clean engine radiator and condenser. Install heavy duty fan or fan shroud, or reposition condenser, whichever is necessary.
	7. Evaporator clogged.	7. Fins clogged with lint, dust, or coated with cigarette tars.	7. Loosen, pull down, and clean with compressed air. Use cleaning solvent to remove cigarette tars.
	Refrigeration		
	1. Too little refrigerant in system.	1. Bubbles in sight glass; high-side gauge readings excessively low.	1. Recharge system until bubbles disappear and gauge readings stabilize to specifications (see "Testing and Adjusting the System" on page 07-1).
	2. Clogged screen in expansion valve.	2. Gauge pressures may be normal or may show slightly low discharge pressure and low suction pressure; discharge output temperature higher than specified.	2. Recover system, remove screen, clean and replace.
	3. Expansion valve thermal bulb has lost charge.	3. Excessively low gauge readings.	3. Recover system; replace expansion valve.
	4. Clogged screen in receiver-dryer.	4. High-side gauge low or normal; low-side gauge lower than normal; receiver-dryer cold to touch and may frost.	4. Recover system; replace receiver-dryer.
	5. Thermostat defective or improperly adjusted.	5. Low-side gauge reading high; clutch cycles at too high a reading.	5. Adjust or replace thermostat.

Diagnosing the System

III. System Cools Intermittently	**Electrical**		
	1. Defective circuit breaker, blower switch, or blower motor.	1. Electrical units operate intermittently.	1. Remove defective part for service or replacement.
	2. Partially open, improper ground, or loose connection in compressor clutch coil or solenoid.	2. Clutch disengages prematurely during operation.	2. Check connections or remove clutch coil or solenoid for service or replacement.
	Mechanical		
	1. Compressor clutch slipping.	1. Visual inspection; compressor operates until head pressure builds up (as viewed on high-side gauge) at which time clutch begins slipping; may or may not be noisy.	1. Slippage over a prolonged period will require that clutch be removed for service; may require readjustment for proper spacing.
	Refrigeration		
	1. Unit icing up may be caused by excessive moisture in system, loose or corroded expansion valve thermal bulb, incorrect superheat adjustment in expansion valve, or thermostat adjusted too low.	1. Unit ices up intermittently.	1. Inspect expansion valve thermal bulb or valve; replace receiver-dryer if excess moisture is present; adjust thermostat.
	2. Thermostat defective.	2. Low-side pressure may be low or excessively high; adjustments will not correct.	2. Replace thermostat.
IV. System Too Noisy	**Electrical**		
	1. Defective winding or improper connection in compressor clutch coil or solenoid.	1. Visual inspection; solenoid or clutch vibrates.	1. Replace or repair as necessary.
	Mechanical		
	1. Loose or excessively worn drive belt(s).	1. Belt(s) slip and are noisy.	1. Tighten or replace as required.
	2. Noisy clutch.	2. May or may not slip; noisy when engaged.	2. Remove clutch for service or replacement as necessary.
	3. Compressor noisy.	3. Loose mountings; worn parts inside compressor.	3. Check mountings and repair; remove compressor for service or replacement.
	4. Loose panels.	4. Excessive rattling during operation.	4. Check and tighten all panels and hose hold-down clamps; check for rubbing or vibrations of hoses or pipes.
	5. Compressor oil level low.	5. Compressor noisy.	5. Discharge system. Fill with specified oil to correct level.
	6. Blower fan noisy; excessive wear in blower motor.	6. Blower motor noisy.	6. Remove blower motor for service or replacement as necessary.
	7. Idler pulley and bearing defective.	7. Whining or growling noise during operation; pulley has rough feel when rotated by hand.	7. Replace bearing; inspect idler and pulley for excessive wear.
	Refrigeration		
	1. Excessive charge in system.	Rumbling noise or vibration in high-pressure line; thumping noise in compressor; excessive head pressure and suction pressure.	1. Discharge excess refrigerant until high-side gauge drops within specifications (see "Condition No. 1" on page 06-8, Inspection Step C).
	2. Low charge in system.	2. Hissing in evaporator case at expansion valve; bubbles or cloudiness in sight glass; low head pressure.	2. Check system for leaks; charge system.
	3. Excessive moisture in system.	3. Suction pressure low.	3. Replace receiver-dryer or accumulator; evacuate and charge system.
	4. High-pressure service valve closed.	4. Compressor has excessive knocking noise; high-side gauge reads above normal.	4. Open valve immediately.

CS33148,000304B -19-08SEP09-4/3

Flow Charts for Diagnosing the System

The following charts are designed to acquaint the air conditioning technician with gauges as diagnostic instruments.

However, before attaching the pressure gauges, make sure other problems that are easier to repair have been checked first. See the discussion on "Troubleshooting Customer Complaints" on page 06-2.

If the gauges must be installed (see "Installing Gauge Set to Check System Operation" on page 07-1), and the problem is not discovered using the charts shown previously, use the following flow charts to determine the cause of a malfunction. Each flow chart begins with an abnormal pressure gauge reading that could be expected if the air conditioning system operates improperly.

After installing the gauges, determine what the normal pressure should be. Normal pressures will vary depending on the ambient temperature (see "Troubleshooting Customer Complaints" on page 06-2 and "Customer Complaints — Troubleshooting Chart" on page 06-3). If pressures are not normal, refer to the appropriate flow chart that matches the gauge readings:

- Condition 1: LOW SIDE — low or vacuum, HIGH SIDE — normal or low.
- Condition 2: LOW SIDE — normal or high, HIGH SIDE — high.
- Condition 3: LOW SIDE — normal, HIGH SIDE — normal; system still not providing sufficient cooling.
- Condition 4: LOW SIDE — high, HIGH SIDE — low.

NOTE: For more specific diagnostic procedures, follow recommendations and specifications from the manufacturer as provided in the Technical Manual.

Diagnosing the System

Condition 1
Low Side: Low or Vacuum
High Side: Normal or Low
Ambient 95°F (35°C)

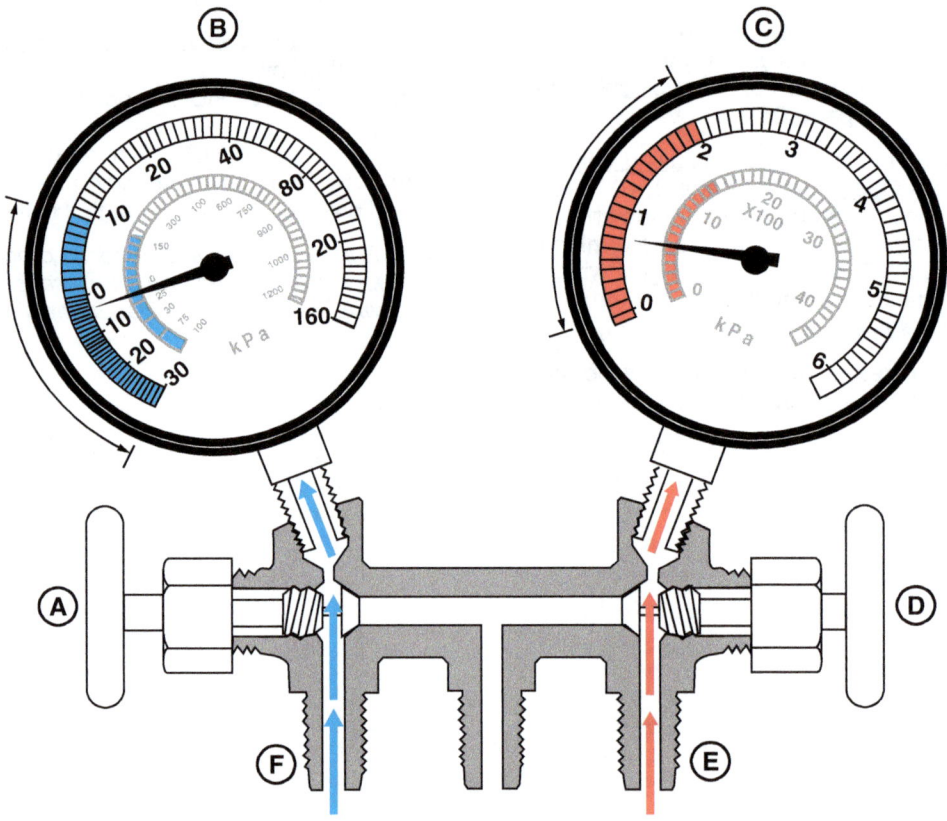

Fig. 2 — Condition 1

A—Low-Side Hand Valve Closed
B—Low Side Low or Vacuum
C—High Side Normal or Low
D—High-Side Hand Valve Closed
E—High-Side Hose Connected to High-Side Service Connector
F—Low-Side Hose Connected to Low-Side Service Connector

Inspection	Repair
Step A	
Inoperative blower motor	Repair blower motor.
Dirty air filter	Clean filter.
Low charge of refrigerant (bubbles in sight glass)	Leak test system. Repair and recharge system. Recheck pressure. If no change, go to Inspection Step B.
Step B	
Check for restrictions: • Between condenser and receiver-dryer • At receiver-dryer • Between receiver-dryer and expansion valve • At expansion valve (inlet screen partially clogged) • Between evaporator and compressor	1. Remove thermal fuse (if so equipped) from clutch lead.
	2. Connect a jumper wire between power and clutch terminal to prevent thermal fuse failure during test.
	3. Replace fuse if it is faulty.
	4. Operate engine at 2000 rpm, with compressor operating for three minutes. Set blower on HIGH.
	5. Check for frost on upper end of expansion valve just before the valve outlet connection. • If frost is present, proceed to step 6. • If no frost, go to Inspection Step C.
	6. Inspect line from condenser to expansion valve for frost or temperature change (temperature change indicates a restriction). • If frost or temperature change is present, repair restriction and recheck pressure. • If no frost or temperature change, go to Inspection Step E.

Continued on next page

Diagnosing the System

Inspection	Repair
Step C	
Inspect evaporator outlet pipe for frost.	If frost is present, go to Repair Step D.
	If no frost is present, inspect the lines between evaporator and compressor for points where frost starts to accumulate or temperature changes. A slight temperature change usually indicates a restriction. • If frost is present or temperature changes, repair restriction and recheck pressures. • If no frost is present or there is no temperature change, go to Inspection Step E.
Step D	
Check pressure readings.	1. With compressor OFF, open vehicle doors for three minutes.
	2. Close doors, start engine (2000 rpm), compressor ON for two minutes. • If pressures are normal, there may be moisture in the system. • If pressure is low, go to Inspection Step E.
Step E	
Check for partially clogged inlet screen on expansion valve for low gas charge in thermal bulb.	Remove expansion valve inlet screen for inspection.
	If screen is dirty: • Clean screen and reinstall. • Flush line between receiver-dryer and expansion valve. • Install a new receiver-dryer. • Add appropriate amount of refrigerant oil.
	If screen is not dirty: • Replace expansion valve. • Do not replace receiver-dryer unless it is more than two years old. • Connect all components and recover system. (See "Preparing System for Service" on page 08-1.) • Evacuate system. (See "Preparing System for Service" on page 08-1.) • Charge system with refrigerant. (See "Preparing System for Service" on page 08-1.)

CS33148,000304D -19-21JAN11-2/2

Diagnosing the System

Condition 2

Low Side: Normal or High
High Side: High
Ambient 95°F (35°C)

NOTE: No bubbles in sight glass.

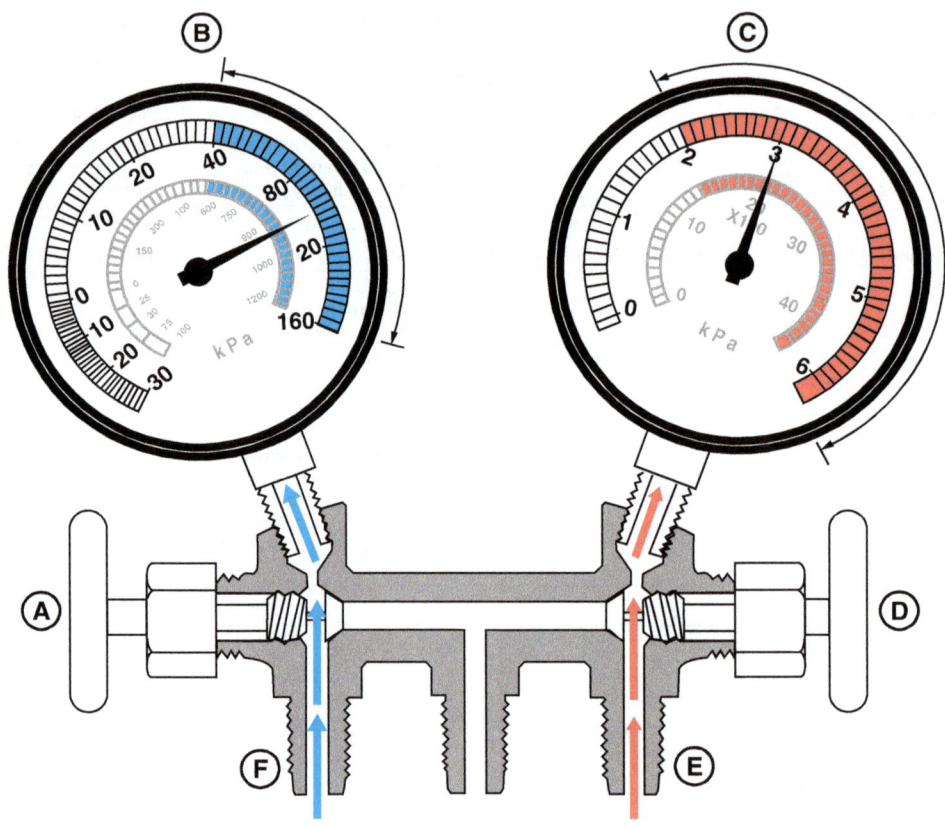

Fig. 3 — Condition 2

A—Low-Side Hand Valve Closed
B—Low Side Normal or High
C—High Side High
D—High-Side Hand Valve Closed
E—High-Side Hose Connected to High-Side Service Connector
F—Low-Side Hose Connected to Low-Side Service Connector

Inspection	Repair
Step A	
Check for restricted air flow through condenser or radiator.	Clean out condenser or radiator. Then check pressure gauges. If no change, go to Inspection Step B.
Step B	
Check thermal expansion valve for loose or corroded thermal bulb.	Repair.
	If not loose or corroded, go to Inspection Step C.
Step C	

Continued on next page

Diagnosing the System

Inspection	Repair
Check for system overcharged with refrigerant.	1. Recover refrigerant in system. (See "Preparing System for Service" on page 08-1.)
	2. Evacuate system. (See "Preparing System for Service" on page 08-1.)
	3. Charge system with refrigerant. (See "Testing and Adjusting the System" on page 07-1.)
	4. Recheck pressures. • If pressures are normal, system is okay. • If high-side pressure is high, open high side manifold gauge valve, adjust pressure, and close valve. Observe low-side pressure gauge for a decrease in pressure. • If pressure did not decrease, go to Inspection Step D. • If pressure did decrease, skip to Inspection Step E.
Step D	
Expansion valve is sticking. Replace expansion valve.	1. Recover system. (See "Preparing System for Service" on page 08-1.)
	2. Remove expansion valve inlet hose and remove screen for inspection.
	3. If screen is dirty: • Clean screen and flush line between receiver-dryer and expansion valve. • Replace receiver-dryer. • Add 0.75 oz (22 mL) of refrigerant oil. (See "Testing and Adjusting the System" on page 07-1.)
	4. If screen is clean, do not replace receiver-dryer unless it is more than two years old.
	5. Install a new expansion valve and connect oil components.
	6. Evacuate system. (See "Preparing System for Service" on page 08-1.)
	7. Charge system with refrigerant. (See "Testing and Adjusting the System" on page 07-1.)
	8. Recheck pressures. (See chart in "Visual Inspection of the System" on page 05-2 and refer to manufacturer's specifications.)
Step E	
Check for expansion valve intermittently sticking open or for air in system.	1. Remove thermal bulb from outlet pipe of evaporator.
	2. Place bulb in palm of hand and close fingers around bulb to warm it for one minute.
	3. Check gauges for increase in pressure.
	4. Reattach bulb after testing or repair.
	5. If high-side pressure is high, open high-side manifold gauge valve, adjust pressure, and close valve. Observe low-side pressure gauge for a decrease in pressure.
	6. If pressure did not decrease, go to Inspection Step F.
	7. If pressure did decrease, go to Inspection Step D.
Step F	
Remove air from the system.	1. Recover refrigerant in system. (See "Preparing System for Service" on page 08-1.)
	2. Evacuate system. (See "Preparing System for Service" on page 08-1.)
	3. Charge system with refrigerant. (See "Testing and Adjusting the System" on page 07-1.)
	4. Recheck pressures.

Diagnosing the System

Condition 3

Low Side: Normal
High Side: Normal
Ambient 95°F (35°C)

NOTE: Although pressures are normal, system still may not provide sufficient cooling.

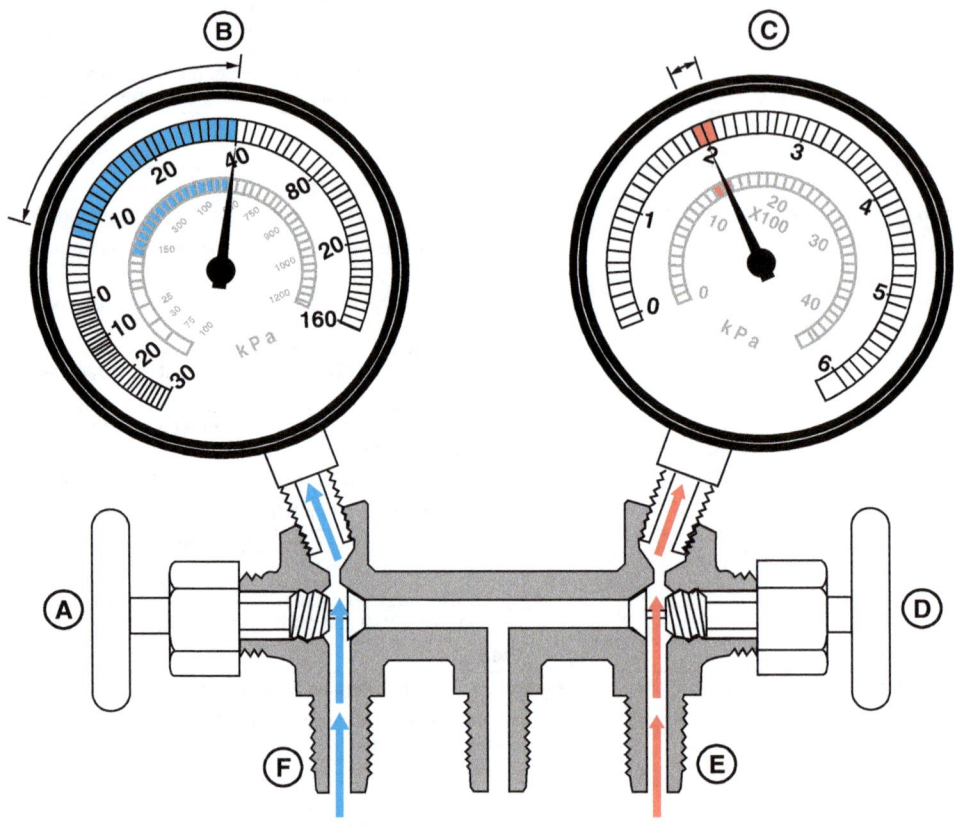

Fig. 4 — Condition 3

A—Low-Side Hand Valve Closed
B—Low Side Normal
C—High Side Normal
D—High-Side Hand Valve Closed
E—High-Side Hose Connected to High-Side Service Connector
F—Low-Side Hose Connected to Low-Side Service Connector

Inspection	Repair
Step A	
1. Run engine at 2000 rpm for at least ten minutes. 2. Operate compressor. 3. Inspect low-side line from evaporator to compressor for frost.	If frost is present, inspect thermal expansion valve thermal bulb. If loose or corroded, repair and then go to Inspection Step B.
	If no frost, go to Inspection Step B.

Continued on next page

Diagnosing the System

Inspection	Repair
	Step B
Check temperature drop at air outlet ducts.	Temperature Drop Check:
	1. Run engine at 2000 rpm and operate compressor.
	2. Put thermometer in blower air duct with blower switch at HIGH.
	3. After at least 20 minutes of operation with doors closed, note air duct temperature.
	4. Compare temperature to Minimum Specification Chart.
	Minimum Specification Chart
	Ambient Air
	Below 75°F (24°C)
	75°–90°F (24–32°C)
	Above 90°F (32°C)
	5. If within specification, go to Inspection Step F.
	6. If not within specification, go to Condition 4.
	7. Use two flat washers to crimp a heater hose shut with locking pliers.
	8. Run engine at 2000 rpm and operate compressor.
	9. Put thermometer in blower air duct with blower switch at HIGH.
	10. After at least 20 minutes of operation with doors closed, note air duct temperature.
	11. Compare temperature to Minimum Specification Chart.
	• If within specification, either the heater valve is leaking internally or the heater hoses are reversed. Repair either condition then repeat temperature drop check again. • If not within specification, go to Inspection Step C.
	Step C
Check doors, windows, and seams for leaks.	Repair any leaks and repeat temperature drop check in Repair Step B.
	Step D
If no leaks are found in Step C, lack of cooling could be caused by dirty components. Check the following: • Recirculating filter • Blower air duct and fan cage • Condenser • Radiator • Evaporator	Clean dirty components and repeat temperature drop check in Repair Step B.
	Step E
If components in Inspection Step D were not dirty, check for temperature change in high-side lines.	1. Run engine at 2000 rpm with compressor operating.
	2. Feel along entire length of high-side line from compressor to expansion valve for change in temperature. Always check for a temperature change in the normal direction of refrigerant flow. High side is normally hot.
	NOTE: Tubing may be dented, kinked, or internally blocked, restricting flow of refrigerant.
	• No temperature change means no restriction — go to Inspection Step F. • If there is a temperature change, repair restriction, then recheck pressures. Refer to chart in "Visual Inspection of the System" on page 05-2.
	Step F
Check clutch cycle time.	See manufacturer's Technical Manual for testing procedure.
	NOTE: If the system checks out to be normal at this step but still there is a lack of cooling, suspect that there is moisture in the system. See "Refrigerants and Oil" on page 02-1 for instructions on removing moisture from system.

Diagnosing the System

Condition 4
Low Side: High
High Side: Low
Ambient 95°F (35°C)

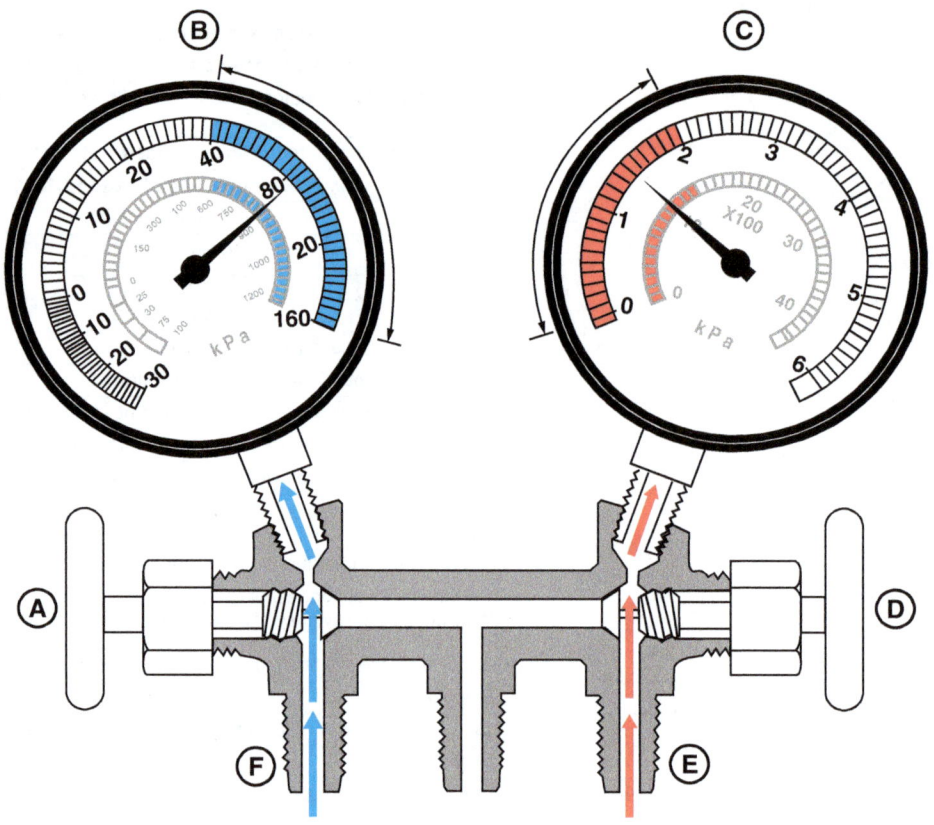

Fig. 5 — Condition 4

A—Low-Side Hand Valve Closed
B—Low Side High
C—High Side Low
D—High-Side Hand Valve Closed
E—High-Side Hose Connected to High-Side Service Connector
F—Low-Side Hose Connected to Low-Side Service Connector

Inspection	Repair
Step A	
Clutch not engaging or belt slipping.	Repair and recheck pressure.
Step B	
Clutch slipping.	Repair.
	NOTE: Repair requires recovery of refrigerant, evacuation of air, and charging.

Diagnosing the System

Diagnostic Chart

The diagnostic chart can help to understand how different conditions can affect the entire system. The chart gives symptoms that normally are associated with the conditions that are listed.

Use the Troubleshooting Charts and the Flow Charts for Diagnosing the System to locate the source of a problem and systematically examine and repair the problem.

Condition	Low-Side Pressure	High-Side Pressure	Sight Glass	Suction Line	Receiver-Dryer	Liquid Line	Discharge Line	Discharge Air
Lack of Refrigerant	Very low	Very low	Clear	Slightly cool	Slightly warm	Slightly warm	Slightly warm	Warm
Loss of Refrigerant	Low	Low	Bubbles	Cool	Warm to hot	Warm	Warm to hot	Slightly cool
Compressor Failure	High	Low	Clear	Cool	Warm	Warm	Warm	Slightly cool
Condenser Malfunction	High	High	Clear to occasional bubbles	Slightly cool to warm	Hot	Hot	Hot	Warm
Expansion Valve Stuck Open	High	High or normal	Clear	Cold — sweating or frosting heavily	Warm	Warm	Hot	Slightly cool
Expansion Valve Stuck Closed	Low	Low	Clear	Cold — sweating or frosting heavily at valve inlet	Warm	Warm	Hot	Slightly cool
Restriction between Condenser and Expansion Valve	Low	Low	Clear	Cool	Cool or sweating or frosting	Cool or sweating or frosting	Hot to point of restriction	Slightly cool
Restriction between Compressor and Condenser	High	High normal or low	Clear	Slightly cool to warm	Warm or hot	Warm or hot	Hot	Warm
Normal	Normal	Normal	Clear	Cool — possible light sweat	Warm	Warm	Hot	Cool (see Minimum Specification Chart in "Condition No. 3" on page 06-12.

CS33148,0003051 -19-26MAR09-1/1

Test Yourself

Questions

1. If the air conditioning system produces no cooling, what are the types of possible causes for the problem and what should be inspected first?

2. List at least four symptoms that would be associated with an expansion valve stuck open.

3. If the expansion valve is sticking open, the low-side pressure would be _____ , and the high-side pressure would be _____ .

See "*Answers to Chapter 6 Questions*" on page B-1.

CS33148,0003052 -19-26MAR09-1/1

Testing and Adjusting the System

Introduction

The following operational checks and adjustments are performed when testing the system and bringing it to best efficiency. These procedures begin with connecting the gauges into the system and carry the service technician through adjustment of various controls. For repairs to the system and actual test readings, refer to the manufacturer's Technical Manual. After repairs, always give the system a performance test and make the necessary adjustments described in this chapter.

Installing Gauge Set to Check System Operation

Follow these steps anytime it becomes necessary to install test gauges into the system. The following service procedures require that the gauges be installed into the system. Use only the steps here that are necessary to perform the particular procedure.

⚠ **CAUTION: Put on a face shield before starting this operation — refrigerant can blind.**

1. Connect service gauge hoses to service connectors (Fig. 1).

    a. Remove high- and low-side gauge port caps.

    b. Connect service gauge hoses to service connectors.

REMEMBER:

- High-side service valve connects to condenser (smaller hose).
- Low-side service valve connects to evaporator (larger hose).
- High-side hose connects below high pressure gauge.
- Low-side hose connects below low pressure gauge.

Fig. 1 — First Identify and Connect Test Hose into High Side of System

A—Low-Side Hose
B—High-Side Hose
C—High-Side Valve
D—Low-Side Valve

Continued on next page

Testing and Adjusting the System

2. Connect refrigerant tank valve to refrigerant recovery and recycling station.

    a. With the compressor off, connect the tank valve to the refrigerant recovery and recycling station. Then open the low- and high-side valves to stabilize the pressure (Fig. 2).

    b. Compare the gauge readings with "R-12 Temperature-Pressure Relation Chart — Low Side" on page 01-9 or "R-134a Temperature-Pressure Relation Chart — Low Side" on page 01-10. Add refrigerant if the pressure is below that specified for ambient temperature. Then close low- and high-side gauge manifold valves.

E—Low-Side Gauge
F—High-Side Gauge
G—Gauge Manifold

Fig. 2 — Open High- and Low-Side Valves to Stabilize Pressure

Adding Refrigerant to the System

A small refrigerant loss between seasons is to be expected and is accepted as normal. When connecting the gauges into the system, the service technician will use the refrigerant recovery and recycling station to purge air from the test hoses (see "Installing Gauge Set to Check System Operation" on page 07-1). The technician will be prepared to add refrigerant should the system require it.

1. Connect manifold hose to refrigerant source.

    a. Gauges connected into system.

    b. System has been operating and is stabilized.

    c. Gauges and/or sight glass indicate shortage of refrigerant.

    d. Connect hose to refrigerant source and bleed air from hose.

2. Add refrigerant to system (Fig. 3).

    a. Air conditioning controls set for maximum cooling and engine operating at the recommended speed, usually 1500–2000 rpm.

    b. Open low-side manifold valve.

IMPORTANT: Be sure that refrigerant enters system only as a VAPOR. Too much liquid entering the compressor can damage internal ports. Be sure to regulate the valve on the container or low-side valve so that the low-side reading will not exceed 40 psi (275 kPa) (2.8 bar). This will ensure that refrigerant in the hose has vaporized before entering the compressor. Also, the fittings on the low-side gauge should feel cold when refrigerant is entering system as a gas.

   c. Open valve to a maximum of 40 psi (275 kPa) (2.8 bar) on low-side gauge.

NOTE: In cool weather [below 80°F (27°C)], heat the refrigerant container to no higher than 125°F (52°C), to help vaporize the refrigerant before it enters the system.

   d. If using 15 oz (0.4 L) cans (only available for use by a certified air conditioning technician), when one is empty, close shutoff valve on dispensing valve and change cans.

3. Check system for full refrigerant charge.

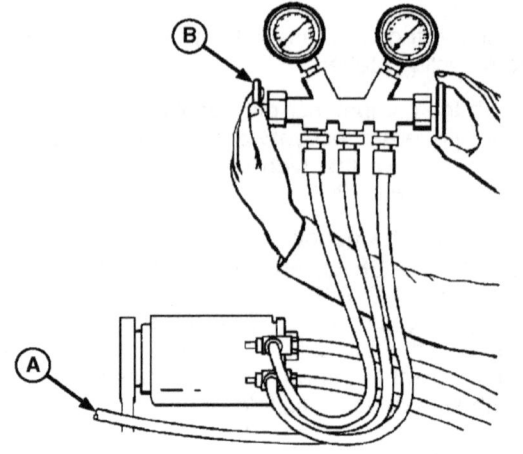

Fig. 3 — Adding Refrigerant with System in Operation

A—To Refrigerant Recovery and Recycling Station
B—Open Low Side

   a. Close low-side hand manifold valve to check for complete charge.

   b. High-side gauge will show normal reading of head pressure in relation to ambient temperature. See "R-12 Temperature-Pressure Relation Chart — Low Side" on page 01-9 or "R-134a Temperature-Pressure Relation Chart — Low Side" on page 01-10.

   c. Low-side gauge will normalize at pressures of 1–30 psi (7–207 kPa) depending on type of control used on system.

   d. Sight glass, if used, will be clear of bubbles. Add additional refrigerant only if specified by manufacturer.

   e. Close valve or refrigerant container.

4. Continue performance test:

   a. Check system for leaks.

   b. Repair system if gauges did not normalize.

   c. Deliver to customer if no further repairs are necessary.

Volumetric Test of Compressor

This test will tell if the compressor is faulty and needs to be replaced.

1. Stabilize system at specified rpm, usually 1500–2000 rpm.

    a. Adjust air conditioning controls for maximum cooling.

    b. Operate for 10–15 minutes.

2. Test the compressor.

    a. Return to idle speed as specified.

    b. Shut off engine.

    c. Leak test the compressor.

3. Recover refrigerant from system.

    a. See "Refrigerant Recovery" on page 08-1.

4. Isolate compressor from system.

    a. Remove refrigerant valves from compressor.

    b. Cover compressor openings immediately with a cover plate to protect O-rings and prevent dirt and moisture from entering.

    c. Plug the open ends of high- and low-side hoses when disconnected from compressor.

5. Remove compressor.

    a. Disconnect clutch coil wires and remove the compressor drive belt.

    b. Drain and measure oil from compressor.

    c. Add back at least 1 oz (30 mL) of oil to the compressor before performing volumetric efficiency test. Tip compressor side-to-side and front-to-rear before mounting in vise.

6. Prepare compressor for testing.

    a. Mount compressor in vise with oil reservoir up.

    b. Inspect outside of compressor for damage.

    c. Rotate clutch hub by hand with low- and high-side ports open. If compressor does not rotate smoothly or easily, replace compressor.

    d. Install a test plate adapter over low- and high-side ports (Fig. 4).

Fig. 4 — Test Plate Adapter Installed

A—Test Plate Adapter

7. Connect pressure gauge hoses to adapter: low pressure gauge hose to discharge fitting, and high pressure hose to other fitting.

NOTE: The low pressure hose is connected to discharge adapter so that pressure readings during testing can be done easily.

   a. Rotate compressor approximately 25 times with both manifold hand valves open.

   b. Close low-side valve and open high-side gauge valve.

   c. Use a socket and speed handle to rotate the compressor clutch hub ten times at a rate of one revolution per second.

   d. Pressure gauge should read 60 psi (415 kPa) (4.1 bar) or higher. A lower reading indicates one or more suction or discharge valves are leaking, an internal leak, or an inoperative valve. The compressor should be replaced under these conditions.

   e. If compressor is serviceable, perform shaft seal leak test.

Shaft Seal Leak Test

1. With adapter test plate and pressure gauges still attached, connect charging hose of manifold to refrigerant recovery and recycling station (Fig. 5).

    a. Open both manifold hand valves.

    b. Open valve on refrigerant container and pressurize compressor to container pressure — 60 psi (414 kPa) (4.1 bar) minimum.

2. Check for leakage from shaft seal and compressor end plate.

    a. If there is end plate leakage, replace the compressor. If there is shaft seal leakage, replace shaft seal assembly.

    b. Close valve on refrigerant container, bleed pressure from compressor, and remove test plate.

    c. Add correct amount of oil to compressor (see "Checking and Adding Oil to Reciprocating Piston Compressors" on page 07-6 or "Checking and Adding Oil to Axial Piston Compressors" on page 07-7). Reinstall the compressor.

IMPORTANT: When compressor is removed from vehicle, do not stand clutch on end or damage may result.

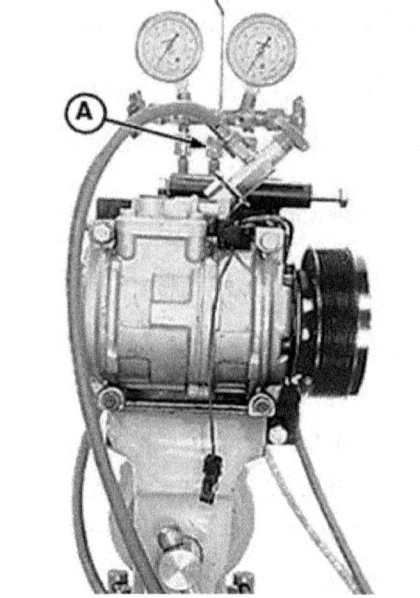

Fig. 5 — Connect Center Hose to Refrigerant Recovery and Recycling Station

A—Center Hose

Checking and Adding Oil to Reciprocating Piston Compressors

This check should be made anytime refrigerant has been added or replaced.

Under normal conditions the oil level need not be checked. There is no place for the oil to go except inside the sealed system. When the engine is first started, some of the oil will be pumped into the rest of the system. After 15 minutes of operation, most of the oil is returned to the compressor crankcase.

Oil circulates with the refrigerant not only to lubricate the compressor, but to lubricate the working parts of the expansion valve as well.

1. Stabilize system at specified rpm, usually 1500–2000 rpm.

    a. Connect gauges into system.

    b. Adjust controls to maximum cooling.

    c. Operate for 10–15 minutes.

2. Isolate compressor from system.

    a. Refer to "Preparing System for Service" on page 08-1.

    OR:

3. Bleed and recover refrigerant from system.

    a. Systems not having service valves must be drained of refrigerant before oil level may be checked.

4. Remove oil check plug.

    a. If service valves are used, bleed compressor charge through low-side manifold hand valve.

    b. Remove cap after low-side gauge reads zero pressure.

5. Check oil level.

    a. Use correct dipstick.

    b. Refer to Technical Manual to determine correct oil level.

    c. Add refrigeration oil of approved viscosity to bring to correct level.

IMPORTANT: Store refrigeration oil in air-tight container to avoid contamination.

6. Place unit in service and continue testing.

    a. Replace oil check plug and purge air from compressor.

    b. Open service valves to mid-position.
    OR:

    c. Evacuate system with suitable vacuum pump.

    d. Charge system with new refrigerant.

Checking and Adding Oil to Axial Piston Compressors

Some axial piston compressors have no provision to accurately check oil level. Periodic inspection of oil level has therefore been eliminated from installations using this compressor. Oil level check, performed with the compressor removed from the vehicle, is to be made in conditions of severe oil loss caused by a compressor seal leak, broken refrigerant hose, or rupture from damage. The check is also necessary when one or more components are replaced and the system is not flushed.

The affinity of refrigerant and refrigeration oil and the design of this compressor will prevent the full oil charge being contained in the compressor. Allow for oil distribution as outlined at the end of this procedure.

1. Bleed and recover refrigerant from system.

    a. Service valves often are not present on installations using this compressor.

    b. Follow "Refrigerant Recovery" on page 08-1 to recover refrigerant.

2. Remove compressor.

    a. Follow procedure as necessary for make and year of model.

3. Remove oil from compressor into clean container.

    a. Remove plug from oil sump.

    b. Pour oil from compressor into container calibrated in ounces (or milliliters) (Fig. 6).

4. Measure oil from compressor.

    a. Determine quantity of oil in compressor.

    b. Refer to machine Technical Manual for amount of oil required to return oil level to full.

NOTE: *When installing a new compressor without flushing the system, drain the compressor to the*

Fig. 6 — Pour Oil into Container

recommended level. This will avoid an overcharge of oil when oil in the new compressor mixes with oil already in the system.

5. Add NEW approved-viscosity oil in compressor (if the system or compressor was recovered).

    a. Properly dispose of old oil removed from compressor.

    b. Clean the container used to measure oil.

    c. Measure required amount of new oil into clean container.

    d. Use small-tipped funnel inserted in drain hole to add oil.

    e. Install plug and tighten.

6. Install compressor.

    a. Follow procedure as necessary for make and year of model.

Continued on next page

7. Add NEW approved-viscosity oil in compressor (if one or more of the components are replaced and the system is not recovered).

   a. With compressor installed on machine, connect gauge manifold hoses to compressor test fittings (Fig. 7).

   b. Put a measured amount of flow oil into the oil injection bottle.

   c. Connect oil injection bottle to suction hose.

   d. Open valve until oil injection bottle has been emptied.

8. Repair damage to system.

   a. Repair system as required to eliminate leak(s).

9. Evacuate system with vacuum pump.

   a. Follow procedure as outlined in "Evacuating System Using Charging Station" on page 08-2.

10. Charge system with NEW refrigerant.

Fig. 7 — Add Oil to Compressor Using Oil Injector Bottle

   a. Follow procedure as outlined in "Charging System Using Charging Station" on page 08-3.

11. Continue performance test.

Testing and Adjusting the System

Bench Testing Expansion Valve for Efficiency

An expansion valve should not be condemned until tested for operating efficiency. Partial blockage in the inlet screen or excessive moisture in the system, which may cause icing near the inlet or outlet of the evaporator, can incorrectly indicate a defective valve. After removing a possible faulty expansion valve, clean the screen and bench test the valve.

Following is an example of a typical expansion valve bench test. Refer to Technical Manual for specific bench test information.

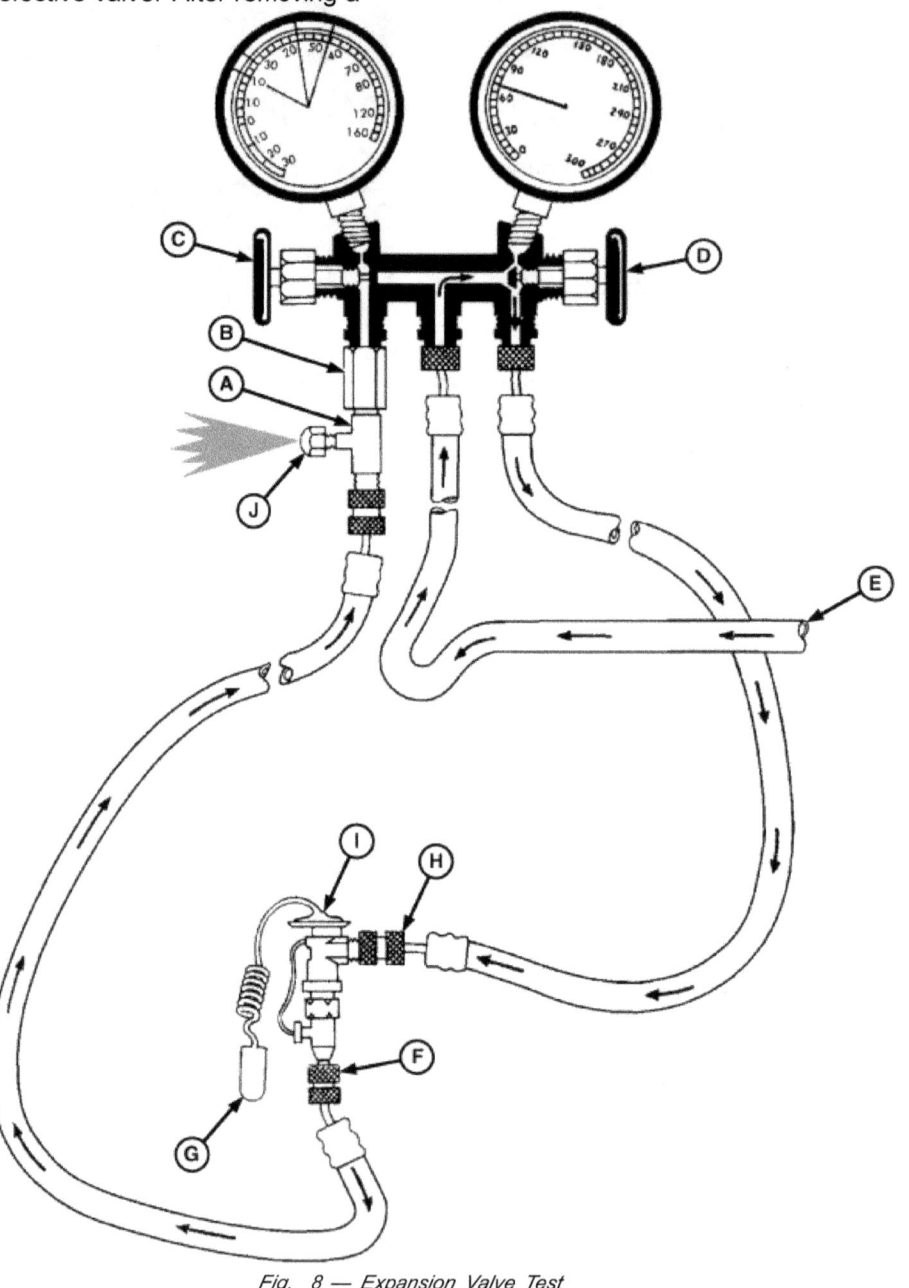

Fig. 8 — Expansion Valve Test

A—1/4-Inch Male Tee
B—1/4-Inch Female Flare Coupler
C—Closed
D—Open
E—Shop Air Supply
F—Outlet
G—Thermal Bulb
H—Inlet
I—Expansion Valve
J—Flare Cap Drilled with No. 71 Drill

1. Install test hose from center connector on gauge manifold to shop air supply (Fig. 8).

   a. Close high- and low-side gauge manifold hand valves.

Continued on next page

Testing and Adjusting the System

b. Install 1/4-inch tee flare fitting to low-side manifold hose connector.

c. Install test hose to lower end of 1/4-inch tee fitting.

d. Install test cap (drilled with No. 71 drill) to side connection on 1/4-inch tee fitting.

e. Install another test hose to high-side manifold hose connector.

2. Connect pressure air to test manifold.

   a. Connect test hose from center connector on gauge manifold to a shop-air line with a nonrestricted blow gun end 1/8-inch male NPT x 7/16-inch-20 (N) flare connector.
   OR:

   b. Prepare test gauges for expansion valve test (Fig. 8).

3. Prepare hot and cold containers. *Use thermometer to obtain near exact temperatures of water used for test.*

   a. Place ice water in a suitable container.

   NOTE: *Add rock salt and stir until a reading of 26–32°F (-3–0°C) is obtained. A cold drink may be substituted for the iced container, provided its temperature is near 28°F (-2°C).*

   b. Heat water in second container until it reaches 115–125°F (46–52°C) or mix hot and cold tap water.

4. Prepare expansion valve for test.

   NOTE: *Always remove screen from expansion valve inlet and clean carefully and reinstall before beginning this test.*

   a. Install 5/8-inch female flare x 1/4-inch male flare to expansion valve outlet and tighten securely.

   b. On expansion valves having a 3/8-inch flare inlet, install 3/8-inch female flare x 1/4-inch male flare reducer and tighten securely.

   NOTE: *Expansion valves used on some systems will require adapters converting O-ring connections to flare fittings. These are available from local refrigeration supply houses.*

   c. Install high-side test hose to inlet fitting on expansion valve.

   d. Install low-side test hose to outlet fitting on expansion valve.

5. Test expansion valve for maximum flow.

   a. Open blow gun valve or, if a refrigerant tank is used, open valve on tank.

   b. Check low-side manifold shutoff valve for closed position.

   c. Place thermal bulb of expansion valve in container of 125°F (52°C) water.

   d. Open high-side gauge manifold hand valve slowly until high-side gauge reads 70–75 psi (483–517 kPa) (4.8–5.2 bar).

   e. Read low-side gauge. It should read 40–55 psi (276–379 kPa) (2.8–3.8 bar).

6. Test expansion valve for minimum flow.

   a. Place thermal bulb in container of 28°F (-2°C) liquid.

   NOTE: *Adjustment of the high-side valve may be required to maintain 70–75 psi (485–515 kPa) (4.8–5.2 bar) inlet pressure.*

   b. Read low-side gauge. It should read 20–25 psi (140–175 kPa) (1.4–1.7 bar).

7. Analyze expansion valve test results.

   - Failure of valve to meet the above test conditions indicates a defective valve.

   a. Replace valve.

8. Install expansion valve into system.

   - An expansion valve that meets above specifications during test has correct superheat setting, valve moving freely, and thermal bulb that has not lost its charge.
   - This expansion valve is suitable for service.

9. Pump down and charge system.

   a. Evacuate system. Refer to "Preparing System for Service" on page 08-1.

   b. Charge system with refrigerant. Refer to "Preparing System for Service" on page 08-1.

10. Continue performance test.

    a. Continue testing system.

    b. Adjust control for maximum performance.

Thermostatic Temperature Control Switch

Some air conditioning systems contain an adjustable thermostatic temperature control switch. A temperature-sensing tube filled with gas is positioned between the refrigerant tubes in the evaporator. It senses the temperature of the evaporator tubes. The gas expands and contracts on a diaphragm that opens and closes electrical contacts in the thermostatic switch.

The switch controls the pumping action of the compressor by means of a magnetic clutch.

If the compressor does not cycle according to Technical Manual specifications, reposition the temperature-sensing tube as specified. The cycle time of the compressor should change.

If unable to obtain specified clutch cycle time, remove the temperature control switch located in the cab. Turn the adjusting screw until the clutch stays engaged for the specified period of time. If still unable to obtain specified cycle time, replace the switch.

Adjusting Thermostat

Thermostat-controlled recycling clutch systems require that the thermostat operation be checked periodically and occasionally adjusted. For thermostat operation, location of capillary tubes, and identification, refer to manufacturer's Technical Manual. The following procedure will detail steps to check a thermostat and how to adjust those that are adjustable.

1. Stabilize system at specified rpm, usually 1500–2000 rpm.

   a. Connect gauges into system with hand valves off.

   b. Adjust air conditioning controls for maximum cooling.

   c. Operate for 10–15 minutes.

2. Read high-side gauge for full refrigerant charge.

   a. Normal high-side pressure will match Pressure-Temperature Table in "Checking Relative Temperatures at High and Low Sides of System" on page 05-3. Refer to manufacturer's Technical Manual.

   b. Check sight glass for absence of bubbles.

3. Adjust air conditioning control for maximum cold.

   a. Thermostat adjusted to coldest position.

   b. Blower fan speed on LOW.

   c. All doors and windows closed.

4. Read low-side gauge for thermostat operation.

   - Should read from 14–26 psi (97–179 kPa) (1.0–1.8 bar) after system is stabilized for 10–15 minutes.

   NOTE: Thermostat should disconnect clutch for evaporator defrost between high and low readings given above. If thermostat will not recycle clutch, move temperature control capillary tube toward warmer position to check for thermostat point opening.

   a. Count number of pounds of pressure required for warm-up until points close — there should be 20–32 psi (138–220 kPa) (1.4–2.2 bar) rise between point opening and reclosing.

   b. Check thermostat operation at least three times for consistent operation.

5. Adjust thermostat.

   NOTE: Thermostats are generally, but not always, located in evaporator case.

   a. Remove parts as necessary to make thermostat accessible.

   b. Open access door to adjustment screw.

   c. Rotate adjustment screw counterclockwise to lower point opening adjustment; clockwise to raise point opening adjustment.

   NOTE: Localities having high mean humidity will require higher point opening than localities with low mean humidity. Coastal areas with point opening adjustment lower than 24–26 psi (165–179 kPa) (1.7–1.8 bar) will result in evaporator freeze-up. Desert areas with very low humidity can easily tolerate point opening adjustment of 14–16 psi (95–110 kPa) (1.0–1.1 bar) without evaporator freeze-up.

   d. Check operation of thermostat for newly adjusted cycle of operation.

   e. Replace thermostat if cycle of operation is inconsistent or will not respond to adjustment.

   f. Replace access door on side of thermostat and any parts removed to reach thermostat.

6. Continue performance test.

   a. Continue testing system.

   b. Remove gauges and deliver machine to customer.

Checking Clutch Coil for Electrical Operation

Use the following procedure as a time-saving device to determine if the coil is defective. Installations may vary; however, the following steps are general enough to meet all requirements.

1. Determine voltage to clutch coil.

NOTE: *With ignition switch ON and clutch energized, battery voltage should be delivered to the coil. To prepare for test, expose connection compressor coil lead for electrical checks.*

⚠ **CAUTION: Do not allow exposed wire to be grounded against machine while switches are in ON position.**

   a. Connect **red** lead of suitable volt-amp tester to exposed wire of connection at clutch coil.

   b. Connect **black** lead of voltmeter to compressor body.

NOTE: *The connections as listed are for a negative-grounded system. Reverse the leads for a positive-grounded system.*

   c. Voltmeter should read battery voltage — if no voltage reading is obtained, check line fuse, voltage to and from ON-OFF switch to locate and repair voltage loss.

2. Determine current draw of clutch coil.

   a. Separate wires at connection of clutch coil.

   b. Connect **red** lead of ammeter to exposed wire of coil lead.

   c. Connect **black** lead of ammeter to clutch coil.

   d. Turn switches to ON position to energize clutch coil.

NOTE: *The connections as listed are for a negative-grounded system. Reverse the leads for a positive-grounded system.*

   e. Ammeter should indicate 3 amps draw for a 12-volt system and 5 amps for a 6-volt system.

   f. Zero amps draw indicates an open circuit inside coil; excessive current draw indicates a short circuit within coil.

3. Determine ground circuit resistance.

NOTE: *Performance of the resistance test requires the current draw of the coil to be within specifications.*

   a. Connect clutch coil lead to coil terminals.

   b. Connect **red** lead of voltmeter to clutch coil ground lead.

   c. Connect **black** lead of voltmeter to battery negative post.

NOTE: *Reverse voltmeter leads for a positive-grounded system.*

   d. Turn switches to ON position to energize clutch coil.

   e. Total resistance from compressor body to battery post cannot exceed 0.3 volt.

   f. If resistance is excessive, clean all connections and metal-to-metal contacts, including engine to frame and compressor to engine, to reduce resistance to specifications.

Testing and Adjusting the System

Leak Testing System Using Electronic Leak Detector

Several types of leak detectors are available. Fig. 9 shows an **electronic leak detector**. Carefully follow manufacturer's instructions when using any leak detector.

1. Stabilize system at specified engine rpm, usually 1500–2000 rpm.

   NOTE: *If system is empty of refrigerant, install a partial charge before continuing.*

   a. Gauges connected into system.

   b. Adjust air conditioning controls for maximum cooling.

   c. Operate for 10–15 minutes.

   d. Shut off machine engine.

   e. Close manifold gauge hand valves.

2. Check system pressure.

   NOTE: *Check low pressure side of system with compressor OFF (the pressure will be higher). Check high pressure side of system with compressor OFF and again with compressor operating.*

   a. A minimum of 50 psi (345 kPa) (3.4 bar) is necessary to detect leaks.

   b. If pressure is too low, open both manifold valves and add refrigerant until adequate pressure is obtained.

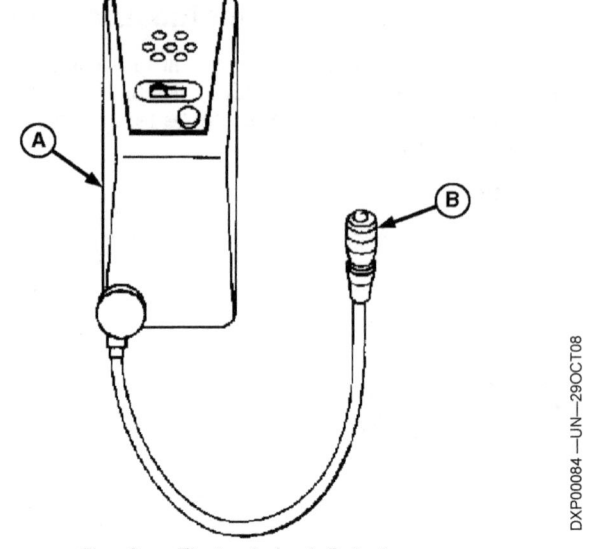

Fig. 9 — Electronic Leak Detector

A—Detector Unit   B—Sampling End

   c. Close both manifold valves and service valve on refrigerant recovery and recycling station.

Continued on next page

Testing and Adjusting the System

3. Move leak detector pickup over system.

   a. Move pickup under hoses, joints seals, and any possible place for a leak to occur (Fig. 10). Do not move sampling end of detector faster than 1 inch (25 mm) per second.

   NOTE: Refrigerant is heavier than air and will move downward; if concentration of refrigerant is located, move pickup slowly upward to locate leak.

   NOTE: Electronic detector registers the presence of refrigerant by a flashing light or a high-pitched squeal.

   b. Repair system as necessary if leaks are located.

   c. Repair, recovery and evacuate the system as required (see "Preparing System for Service" on page 08-1).

4. Start machine and turn compressor ON.

   a. Repeat checks for leaks on high side only.

   ⚠ **CAUTION: When engine is running, be alert and stay clear of rotating parts.**

   - Damp, dusty spots indicate a refrigerant leak.

   b. Repair system as necessary if leaks are located.

5. Check sensitivity of detector pick-up.

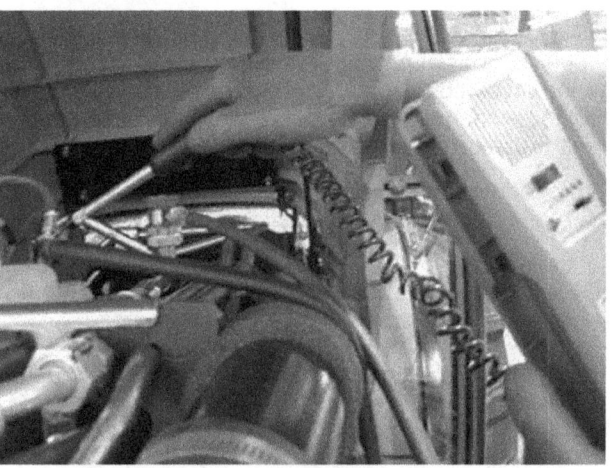

Fig. 10 — Leak Test System Using Electronic Detector

   a. Pass pick-up hose over empty can.
   OR:
   b. Crack open refrigerant container.
   - Electronic unit should indicate by light blinking or by squeal.
   - If no reaction, leak detector is malfunctioning.

6. Resume performance test.
   - Charge system if repairs to system were necessary. See "Preparing System for Service" on page 08-1.

Test Yourself

Questions

1. Fill in the blanks with component names. "The high-side service valve on the compressor leads to the _____ . The low-side valve leads from the _____ ."

2. When inspecting the system, which side is normally warm to hot, and which side is sweating or frosting?

3. True or false? "A slight refrigerant loss between seasons is accepted as normal for an air conditioning system."

4. When adding refrigerant to the system, must it enter as a vapor or as a liquid?

5. When using an electronic leak detector, what does a blinking light or squealing noise tell about the presence or absence of refrigerant?

See "Answers to Chapter 7 Questions" on page B-2.

Preparing System for Service

Introduction

A number of special procedures are required when preparing an air conditioning system for repair service and placing the system back into operation. This chapter will address these procedures.

Each procedure is detailed separately, beginning with recovery of refrigerant from the system, then evacuation for moisture removal, charging the system, and isolating the compressor for service. These procedures are necessary for satisfactory system performance.

Refer to the Technical Manual for the system to get the exact specifications for each procedure.

Refrigerant Recovery

1. Run air conditioning system for three minutes to help in the recovery process (Fig. 1). Turn air conditioning system off before proceeding with recovery steps.

2. Connect discharge hose from manifold set to a refrigerant recovery system or to a refrigerant recovery and recycling system.

3. Open both valves of gauge set. Make certain gas and liquid valves on the refrigerant recovery and recycling tank are open.

4. Close accumulator and oil drain valves.

5. Turn on the main power switch on the refrigerant recovery and recycling system.

6. Be sure the recycle start switch is in the OFF position.

7. Depress the recovery start switch. The amber ON light will come on and the compressor will start. The compressor will shut off automatically when recovery is complete. Wait for five minutes and watch for a pressure rise above 0 psi. Repeat this step if pressure rises above 0 psi (0 kPa) (0 bar). Maintain 0 psi (0 kPa) (0 bar) for two minutes.

8. Drain the oil separator by opening the accumulator pressurizing valve long enough to allow some compressor discharge pressure back into the separator. Open the oil drain valve slowly and drain oil into the oil catch bottle. When the recovered oil has been completely drained, close the valve.

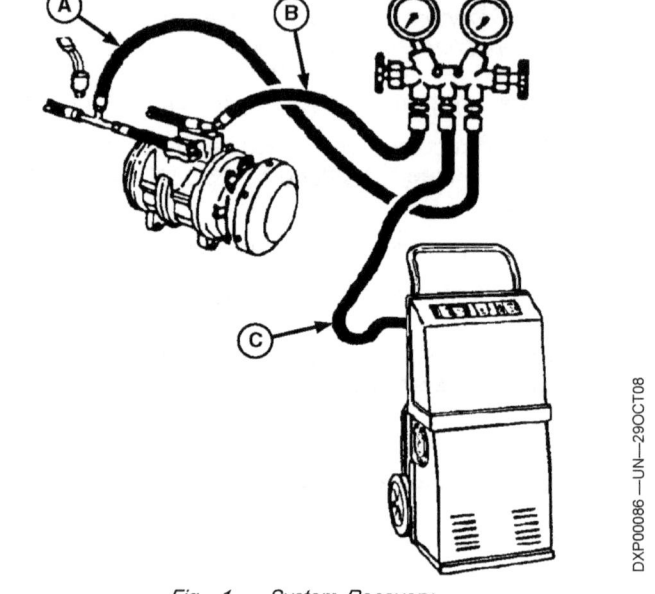

Fig. 1 — System Recovery

A—High-Pressure Hose
B—Low-Pressure Hose
C—Discharge Hose

9. The oil lost during the recovery process must be replaced with new oil as part of system recharging. Measure the amount of oil in the oil catch bottle and add the same amount of new compressor oil to the system.

Evacuating System Using a Charging Station

A vacuum pump is built into the **charging station** (Fig. 2) and is constructed to withstand hard use without damage. Complete moisture removal from the system is possible only with a vacuum pump constructed for this purpose.

1. Operate vacuum pump.

   a. Connect hose to vacuum pump if system was purged through station.

   b. Open high- and low-side gauge valves on charging station.

   c. Connect station into 110-volt current.

   d. Engage ON-OFF switch to vacuum pump according to directions of specific station being used.

   NOTE: *System should pump down into a 28–29-1/2 inch (95–100 kPa) vacuum in not more than 5 minutes. If system fails to meet this specification, repair is necessary.*

   e. Operate pump at least 30 minutes for air and moisture removal.

2. Close hand valves.

   a. Close high- and low-side gauge valves on charging station.

   b. Open switch to turn off vacuum pump.

3. Check ability of system to hold vacuum.

   a. Watch compound gauge to see that gauge does not rise at a rate faster than 1 inch Hg (3.4 kPa) every 4 or 5 minutes.

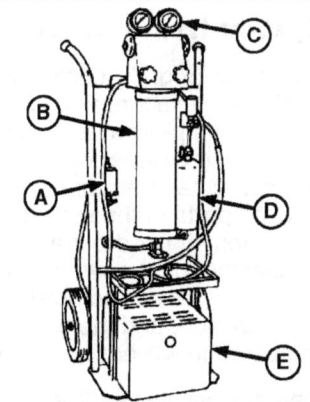

Fig. 2 — Portable Charging Station

A—Oil Injector Cylinder
B—Charging Cylinder
C—Gauges
D—Leak Detector
E—Vacuum Pump

   b. If rise rate of compound gauge is not within specifications, repair system as necessary.
   OR:

   c. If rise rate is within specified time, continue with step 4.

4. Charge system with refrigerant.

   a. Follow steps in "Charging System Using Charging Station" on page 08-3.

Charging System Using Charging Station

Most charging stations (Fig. 3) contain a **charging cylinder** into which the exact amount of refrigerant is added while system pump-down is being performed. The refrigerant charging cylinder contained in the station is heated to the correct temperature to ensure proper refrigerant flow to all parts of the system as a gas during the charging operation. If used correctly, the vacuum pump will so efficiently pump down the system that opening the correct valves will completely charge the system from the high side, and the use of the compressor in the charging operation will not be required.

1. Prepare charging cylinder for filling:

    a. Open refrigerant storage drum valve.

    b. Close all valves on station.

    c. Read storage tank gauge pressure.

    d. Rotate dial shroud on charging cylinder to correlate with pressure on gauge.

    e. Open cylinder fill valve.

2. Fill charging cylinder.

    a. Determine system capacity using Technical Manual.

    b. Intermittently open and close pressure relief valve.

NOTE: *When pressure relief valve opens, refrigerant will enter cylinder and boil. Closing the valve will increase pressure on refrigerant, changing it to liquid to stabilize the refrigerant in the sight glass.*

    c. Fill to specified level in sight glass.

    d. Close pressure relief valve.

3. Charge system with refrigerant.

    a. Connect gauges into system.

    b. Open refrigerant control valve.

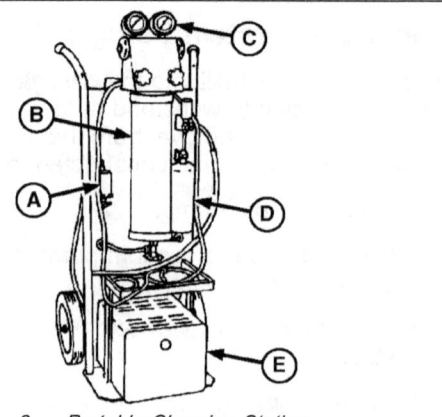

Fig. 3 — Portable Charging Station

A—Oil Injector Cylinder
B—Charging Cylinder
C—Gauges
D—Leak Detector
E—Vacuum Pump

    c. Open high-pressure valve.

    d. Remove vacuum hose from pump and crack open (barely open) low pressure valve.

    e. Allow refrigerant to escape through vacuum hose for approximately 3 seconds.

    f. Close high- and low-pressure valves.

    g. Close refrigerant control valve.

NOTE: *Charging cylinder should empty in approximately 90 seconds for systems with a 5-pound (2.3 kg) capacity. Smaller systems will require less time.*

4. Performance test system.

    a. Continue testing system.

    b. Adjust controls for maximum efficiency.

Charging System Using 15 Ounce (0.4 L) Containers

IMPORTANT: **Only a certified air conditioning technician is authorized to use a refrigerant container smaller than 20 lb (8 L). This system should be charged with refrigerant only after it has been leak tested and evacuated. It is important to add only the specified quantity of refrigerant.**

The tendency of many service technicians is to unknowingly overfill the system. To aid in more accurate charging and to prevent waste, refrigerant manufacturers package the refrigerant in cans that contain 15 ounces (0.4 L); however, the technician must be certified to service systems using containers smaller than 20 lb (9.1 kg). They are handled in the same manner as the larger drums except care must be taken not to overheat the cans because they may explode. It is recommended that a charging station be used to charge a system.

NOTE: *If only a small amount of refrigerant must be added to the system, see "Adding Refrigerant to the System" on page 7-3.*

1. Install can dispensing valve to container(s).

NOTE: *Engine must be OFF. System must be holding vacuum as specified in Step 4 of "Evacuating System Using Vacuum Pump" on page 8-1 or Step 3 of "Evacuating System Using a Charging Station" on page 8-2.*

NOTE: *The dispensing valve is available both for single cans and multiple cans. Whichever is used, preliminary installation to the can(s) is the same.*

   a. Install dispensing valve to single container of refrigerant.
   b. Close shutoff valve on dispensing valve.
   c. Pierce can with mechanism that is part of valve.

2. Install charging hose to dispensing valve (Fig. 4).

NOTE: *Before charging, the system will have been pumped down.*

   a. Loosen charging hose at center connector on gauge manifold.
   b. Crack open dispensing shutoff valve to purge air from charging hose.
   c. Tighten charging hose connection or gauge manifold and close shutoff valve.

3. Partially charge system.

   a. Open shutoff valve on dispensing valve.
   b. Open high side gauge manifold hand valve.

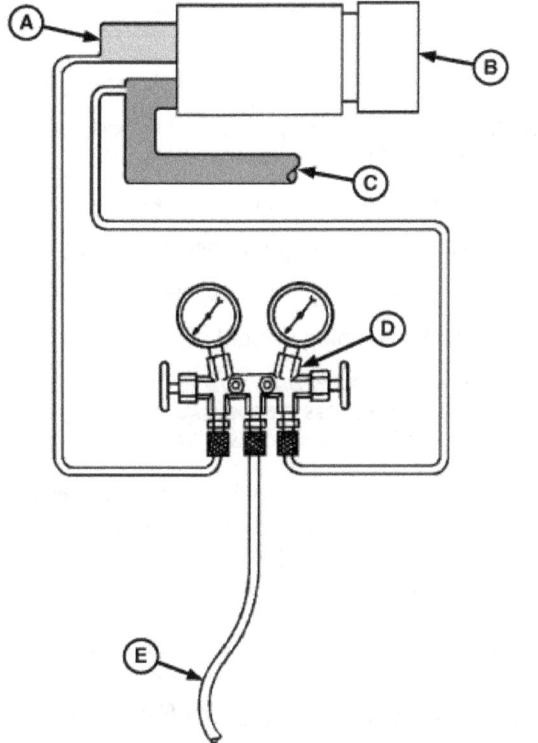

Fig. 4 — Typical Service Valve on Compressor

A—From Evaporator
B—Compressor
C—To Condenser
D—Pressure Gauge Set
E—Charging Hose to Refrigerant Tank

   c. Invert container(s) to allow refrigerant to enter high side of system (Fig. 4).
   d. Close shutoff valve when changing cans.

4. Complete charging of the system.

   a. After high-side pressure becomes slow to increase, open low-side manifold hand valve.
   b. After low-side pressure becomes slow to increase, close high-side manifold valve.
   c. Close shutoff valve on dispensing valve.
   d. Start engine and adjust throttle to specified rpm, usually 1500–2000 rpm.
   e. Adjust air conditioning controls for maximum cooling and engage compressor clutch.
   f. Open shutoff valve on dispensing valve to allow refrigerant to be drawn into system. Adjust low-side pressure valve so gauge reading does not exceed 40 psi (276 kPa) (2.8 bar).

IMPORTANT: **To add refrigerant, see "Adding Refrigerant to the System" on page 7-3.**

NOTE: *If single containers are used, it will be necessary to replace each as it becomes empty.*

Continued on next page

Preparing System for Service

g. Watch sight glass (Fig. 5) until bubbles disappear.

h. Add additional refrigerant only if recommended by manufacturer. See Technical Manual.

i. Close valve on refrigerant container.

j. Close low-side gauge manifold hand valve.

5. Check refrigerant charge in system.

a. Watch for bubbles in sight glass (if used in system).

NOTE: *Excessive head pressure with a normal low-side pressure indicates an overcharge of refrigerant or air in the system. Compressor may or may not be noisy.*

b. Listen for hissing noise in expansion valve. Many systems have a hissing in the expansion valve until the system is fully charged.

6. Continue performance test.

Fig. 5 — Add Refrigerant Until Gauges Normalize and Bubbles in Sight Glass Disappear

F—Sight Glass

a. Continue testing system.

b. Adjust controls for maximum efficiency.

Isolating Compressor from System

On systems having both high-side and low-side service valves, the compressor may be isolated and refrigerant retained in the system while service work is being performed on the compressor or the machine engine. The following procedure should be followed anytime compressor isolation is required.

NOTE: *In some systems the compressor cannot be isolated from the system. In this case, the system must be recovered whenever the compressor is removed.*

1. Stabilize system at specified rpm, usually 1500–2000 rpm.

a. Connect gauges into system.

b. Adjust air conditioning controls for maximum cooling.

c. Operate system for 10 to 15 minutes.

2. Isolate compressor.

a. Slowly close (front-seat) low-side service valve until low-side gauge reads zero pressure (see "Service Valves" on page 04-6 for proper service valve position).

NOTE: *Return engine to idle to prevent "dieseling."*

b. Turn off machine engine.

c. Completely close low-side service valve.

d. Close high-side service valve.

e. Recover refrigerant from compressor by cracking low-side hand manifold until both gauges read zero pressure.

NOTE: *Recover refrigerant slowly to prevent pulling oil from compressor.*

3. Continue service work.

a. Remove service gauges from service valves.

b. Remove service valves from compressor.

c. Perform service work as required.

4. Place compressor in system.

a. Install service valves to compressor using new gaskets or O-rings, whichever are required.

b. Purge air from compressor by cracking low-side service valve for 3 seconds with low-side hose connector capped and high-side hose connector open.

5. Continue performance test.

a. Install gauges to service valve connectors and purge air from hoses.

b. Mid-position service valves.

c. Continue testing system.

d. Adjust control for maximum performance.

Test Yourself

Questions

1. What procedure must be performed before an operable system is disconnected for service?

2. What happens if refrigerant is recovered from system too fast?

3. Why must the system be evacuated of all air before placing it back into service?

4. What three procedures should be carried out before charging a system with refrigerant?

5. If the system has a sight glass, what do bubbles or foamy refrigerant tell you?

6. What option must the compressor have if it is to be isolated from the system while it is being serviced?

See *"Answers to Chapter 8 Questions"* on page B-2.

Definitions and Conversions

Definitions of Terms and Symbols

A

ABSOLUTE ZERO—Complete absence of heat; believed to be −459.67°F (−273.15°C).

AIR CONDITIONING—Absolute control of temperature and humidity; air conditioning in true sense used only in some laboratories and manufacturing plants where temperature and humidity control are very critical. Ordinary usage in homes, buildings, and vehicles means control of temperature and removal of moisture by condensation; more correct designation is refrigeration.

AMBIENT TEMPERATURE—Temperature of surrounding air. In air conditioning, it refers to outside air temperature.

ATMOSPHERIC PRESSURE—Weight of air at various altitudes. Sea level pressure commonly called 14.7 psi (100 kPa) and decreases with higher altitude. Greatest concentration of population of United States lives at 900 feet (275 meters) altitude or less. Society of Automotive Engineers (SAE) uses 900 feet (275 meters) altitude as average for specifications in manufacturing of all products; it is the "mean" or average altitude.

B

BOILING POINT—Temperature at which a liquid changes to a vapor. Water changes to steam at 212°F (100°C) at sea level, 14.7 psi (100 kPa) atmospheric pressure. R-12 changes from liquid to vapor at −22°F (−30°C) sea level and atmospheric pressure. R-134a changes from liquid to vapor at −15°F (−26°C) sea level and atmospheric pressure.

Btu—Abbreviation for British thermal unit. Amount of heat required to raise temperature of one pound of water 1°F (0.56°C). All substances are rated in relation to water as standard of measurement.

C

CELSIUS—Thermometer scale based on 0°C as freezing point of water and 100°C as the boiling point.

CHARGE—Specific amount of refrigerant by weight or volume.

COMPRESSION—Reduction in volume and increase of pressure of a gas or vapor.

COMPRESSOR—Component used to change low-pressure refrigerant to high-pressure refrigerant.

COMPRESSOR CLUTCH—An electromagnetic coupling that engages or disengages the belt-driven compressor pulley to the compressor shaft.

COMPRESSOR DISPLACEMENT—Figure obtained by multiplying displacement of compressor cylinder or cylinders by a given rpm, usually average engine speed.

COMPRESSOR SHAFT SEAL—A seal surrounding the compressor shaft that permits the shaft to turn without loss of refrigerant or oil.

CONDENSATION—Process of changing a gas to a liquid.

CONDENSER—Radiator-type component where refrigerant gives off heat by being changed from a gas to a liquid.

CONDENSING TEMPERATURE—Temperature at which compressed gas in condenser changes from gas to a liquid. Affected directly by quantity and temperature of ram air passing through condenser.

CONDENSING PRESSURE—Head pressure as read from gauge at high-side service valve; pressure from discharge side of compressor into condenser.

CONDUCTION OF HEAT—Ability of substance to conduct heat (metal and glass conduct heat more readily than insulating material such as spun glass).

D

DENSITY—Weight or mass of a gas, liquid, or solid.

DESICCANT—A drying agent used inside air conditioning systems to absorb and hold moisture. Silica gel or molecular sieve are most widely used.

DISCHARGE LINE—Line connecting compressor outlet to the condenser inlet.

DRYER—A device containing a desiccant placed in series, usually in the liquid line, to absorb and hold excess moisture. Usually called receiver-dryer.

E

EVACUATE—To create a vacuum to remove air and moisture in the system.

EVAPORATION—Process of changing a liquid to a gas.

EVAPORATOR—Component where liquid refrigerant is changed to a gas as it absorbs heat from inside air.

EXPANSION—Reduced pressure on liquid refrigerant lowers boiling point, and refrigerant changes to a gas and absorbs heat.

EXPANSION VALVE—Device which restricts flow of high-pressure refrigerant thus lowering refrigerant pressure.

F

FAHRENHEIT—Thermometer scale based on 32°F as freezing point of water and 212°F as the boiling point of water.

FILTER—A device used with the dryer or as a separate unit to remove foreign substances from the refrigerant. Installed in series in liquid line on high side of system.

FLOODED EVAPORATOR COIL—Too much liquid refrigerant in evaporator coil, resulting in poor cooling.

G

GAS—A vapor having no particles or droplets of liquid.

Continued on next page

Definitions and Conversions

GAUGE SET—A set of gauges attached to the compressor service valves for testing or measuring pressure or vacuum.

H

HEAD PRESSURE—Pressure of refrigerant from discharge reed valve through lines and condenser to expansion valve orifice.

Hg—Chemical symbol for mercury. Inches of mercury is a measure of pressure or vacuum.

HIGH SIDE—Same as head pressure; side of system which includes vapor into condenser and liquid into expansion valve. (Also see LOW SIDE.)

HOT GAS BYPASS LINE—Line connecting compressor outlet to evaporator inlet.

HYDROLYZING ACTION—Corrosive action within the air conditioning system. It is induced by a weak solution of hydrochloric acid formed by excessive moisture in the system reacting chemically with the refrigerant.

J

JOULE—In air conditioning it is the metric unit of energy or heat (1 Btu = 1055 J)

K

kPa—Symbol for kilopascal, which is the metric measure of pressure (1 psi = 6.895 kPa).

L

LATENT HEAT—Amount of heat energy required to change a substance from one state of matter to another without changing its temperature.

LATENT HEAT OF CONDENSATION—Quantity of heat energy given off while changing a substance from a vapor to a liquid.

LATENT HEAT OF FREEZING—Heat given off as a liquid is changed to a solid.

LATENT HEAT OF LIQUIDATION—Heat that must be added to change a solid to a liquid.

LATENT HEAT OF VAPORIZATION—Quantity of heat energy required to change a liquid into a vapor without raising temperature of vapor above that of original liquid.

LIQUID LINE—Pipe or hose connecting condenser to expansion valve.

LOW SIDE—That portion of a system from orifice in expansion valve through evaporator line or lines through compressor service valve to compressor reed valve. Also called suction side. (Also see HIGH SIDE.)

M

MODULATOR VALVE—Device which limits and maintains minimum pressure in the evaporator.

P

PRESSURE—Force upon a body, as force upon a liquid, increases the liquid's boiling point.

PRESSURE DROP—Difference in pressure between any two points caused by friction, restriction, etc.

PSI—Abbreviation for pounds per square inch above atmospheric pressure. "G" added designates gauge pressure, but in most applications the G is assumed.

R

RADIATION—Heat flow through space, traveling and acting much like light rays.

RAM AIR—Air that is forced around the condenser coils as the vehicle travels in a forward direction.

RECEIVER-DRYER—See DRYER.

RECOVERY EQUIPMENT—Usually a mechanical system that consists of an evaporator, oil separator, compressor, and a condenser, which draws refrigerant out of a refrigeration system and stores it in a container.

REFRIGERANT—Liquid used in refrigeration system that produces cold by removing heat.

S

SCHRADER VALVE—Spring-loaded valve similar to the tire valve, located inside the gauge hose fitting on service valves and certain controls. Will hold refrigerant in the system but can be opened by installing a special adapter with the gauge hose.

SENSIBLE HEAT—Heat which causes a change in temperature of a substance, but not a change in state.

SIGHT GLASS—Window in receiver-dryer or in liquid line to observe refrigerant flow.

SPECIFIC HEAT—Quantity of heat required to change the temperature of some amount of a substance.

STANDARD TON—Amount of heat released while changing one ton of 33°F (1°C) water to 32°F (0°C) ice in a period of 24 hours. 288,000 Btu (304 MJ) per 24 hours or 12,000 Btu (12.7 MJ) per hour.

STARVED EVAPORATOR COIL—Not enough refrigerant supplied to the coil, resulting in poor operation and a too-low heat exchange.

SUBSTANCE—Any form of matter that can be weighed or measured; may be solid, liquid, or gas.

SUCTION LINE—Line connecting evaporator outlet to compressor inlet.

SUCTION SIDE—Low-side pressure (from expansion valve orifice to intake reed valve in compressor).

SUCTION THROTTLING—Control used to regulate flow of refrigerant from the evaporator to condenser.

Continued on next page

Definitions and Conversions

SUPERHEAT—Added heat intensity to a gas after complete evaporation of a liquid; controlled by increasing pressure in air conditioning systems.

T

TAIL PIPE—Outlet pipe from evaporator coil.

TOTAL HEAT LOAD—Human heat load plus heat entering through floor, glass, roof, and sides of vehicle.

TORQUE—Rotating power required to properly tighten a bolt or nut expressed in pound-feet or pound-inches (newton-meters).

V

VACUUM—Referred to as less than atmospheric pressure and expressed as inches of mercury in Hg or kilopascals (kPa).

VISCOSITY—The measure of resistance of a fluid to flow.

W

WATT—In air conditioning, capacity is shown in Btu or watts (1 Btu/hr = 0.293 W/hr).

Definitions and Conversions

Measurement Conversion Chart

Metric to English

LENGTH
1 millimeter = 0.03937 inches .... in.
1 meter = 3.281 feet.. ft
1 kilometer = 0.621 miles .... mi

AREA
1 meter^2 = 10.76 feet^2 .... ft^2
1 hectare = 2.471 acres .... acre
(1 hectare = 10,000 m^2)

MASS (WEIGHT)
1 kilogram = 2.205 pounds .... lb
1 tonne (1000 kg) = 1.102 short ton .... sh tn

VOLUME
1 meter^3 = 35.31 foot^3 .... ft^3
1 meter^3 = 1.308 yard^3 .... yd^3
1 meter^3 = 28.38 bushel .... bu
1 liter = 0.02838 bushel .... bu
1 liter = 1.057 quart.. qt

PRESSURE
1 kilopascal = 0.145 pound/inch^2 .... psi

STRESS
1 megapascal or
1 newton/millimeter^2 = 145 pound/inch^2 .... psi
(1 N/mm^2 = 1 MPa)

POWER
1 kilowatt = 1.341 horsepower (550 lb-ft/s) .... hp
(1 J = 1 watt/s)

ENERGY (WORK)
1 joule = 0.0009478 British Thermal Unit .... Btu
(1 J = 1 watt/s)

FORCE
1 newton = 0.2248 pounds-force .... lb force

TORQUE OR BENDING MOMENT
1 newton meter = 0.7376 foot-pound .... lb-ft

TEMPERATURE
$t_C = (t_F - 32)/1.8$

English to Metric

LENGTH
1 inch = 25.4 millimeters... mm
1 foot = 0.3048 meters.. m
1 yard = 0.9144 meters... m
1 mile = 1.608 kilometers.. km

AREA
1 foot^2 = 0.0929 meter^2 .... m^2
1 acre = 0.4047 hectare .... ha
(1 hectare = 10,000 m^2)

MASS (WEIGHT)
1 pound = 0.4535 kilograms .... kg
1 ton (2000 lb) = 0.9071 tonnes... t

VOLUME
1 foot^3 = 0.02832 meter^3 .... m^3
1 yard^3 = 0.7646 meter^3 .... m^3
1 bushel = 0.03524 meter^3 .... m^3
1 bushel = 35.24 liter L
1 quart = 0.9464 liter L
1 gallon = 3.785 liter... L

PRESSURE
1 pound/inch^2 = 6.895 kilopascals.. kPa
1 pound/inch^2 = 0.06895 bar .... bar

STRESS
1 pound/inch^2 (psi) = 0.006895 megapascal .... MPa
or newton/mm^2 .... N/mm^2
(1 N/mm^2 = 1 MPa)

POWER
1 horsepower (550 lb-ft/s) = 0.7457 kilowatt........................... kW
(1 watt = 1 N·m/s)

ENERGY (WORK)
1 British thermal unit = 1055 joules .... J
(1 watt = 1 N·m/s)

FORCE
1 pound-force = 4.448 newtons .... N

TORQUE OR BENDING MOMENT
1 pound-foot = 1.356 newton-meters .... N·m

TEMPERATURE
$t_F = 1.8 \times t_C + 32$

Answers to Test Yourself Questions

Answers to Chapter 1 Questions

1. **Absorbs** heat.
2. Raises the boiling point.
3. First blank—**hotter**; second blank—**colder**.
4. The amount of heat required to raise the temperature of a pound of water one degree Fahrenheit (at sea level pressure).
5. Refrigerant-134a does not have a chlorofluorocarbon (CFC) base as Refrigerant-12 has. Therefore R-134a is not as harmful to the environment.
6. See the complete diagram in Fig. 11 in the text.

CS33148,000306A -19-26MAR09-1/1

Answers to Chapter 2 Questions

1. False. It will boil.
2. Frostbite can occur.
3. It can explode.
4. It causes corrosion of the metal parts.
5. A vacuum pump.
6. False! Never use motor oil; use only approved refrigeration oil.
7. Polyalkalene Glycol (PAG).

CS33148,000306B -19-26MAR09-1/1

Answers to Chapter 3 Questions

1. See completed diagram in Fig. 1 of the text for correct labels for parts and colors for refrigerant.
2. The second job is **circulating** the refrigerant in the system.
3. **Concentrates** its heat content or **heats it up**.
4. First blank—**heat;** second blank—**gas;** third blank—**liquid.**
5. **Ram air** is natural air flow from vehicle movement; **forced air** is air pushed by an electric-powered fan.
6. The **outlet** side should be cold.
7. First blank—**liquid;** second blank—**gas;** third blank—**absorbed.**
8. At its **lowest** speed to allow the greatest absorption of heat from the air.
9. To absorb moisture from the system.

CS33148,000306C -19-26MAR09-1/1

Answers to Chapter 4 Questions

1. Gauge and manifold set, leak detector, refrigerant recovery and recycling station.
2. One for the high side, and one for the low side of the system.
3. The compressor.
4. The electronic type.
5. It gives off a poisonous gas.
6. A vacuum pump.

CS33148,000306D -19-26MAR09-1/1

Answers to Chapter 5 Questions

1. Normal is 1–80 psi (50–210 kPa) (depending upon ambient temperature).
2. See "Pressure-Temperature Table" on page 05-3 for correct answers.
3. 1-b; 2-c; 3-a.

CS33148,000306E -19-26MAR09-1/1

Answers to Chapter 6 Questions

1. The possible causes that take the least amount of time to inspect and repair.
2. See "Diagnostic Chart" on page 06-14.
3. First blank—**High**; second blank—**High or normal.**

CS33148,000306F -19-26MAR09-1/1

Answers to Test Yourself Questions

Answers to Chapter 7 Questions

1. First blank—**condenser;** second blank—**evaporator.**
2. The high side will be warm to hot; the low side will be sweating or frosting.
3. True.
4. As a vapor.
5. Blinking light or squealing noise indicates the presence of refrigerant.

Answers to Chapter 8 Questions

1. The system must be **recovered** to release pressure.
2. Too-rapid recovering will **draw out too much oil** with the refrigerant from the compressor and system.
3. To **remove moisture,** which contaminates the system and corrodes its working parts.
4. Leak test; purge; evacuate.
5. That the **refrigerant is low.**
6. It must have high- and low-side service valves.

B-2

Index

A

Accumulators 03-17
Adding Refrigerant to the System 07-3
Adjusting Thermostat 07-11

B

Bench Testing Expansion Valve for 07-9
Bypass Systems 03-22

C

Chapter Introductions 01-1, 02-1, 03-1, 04-1, 05-1, 06-1, 07-1, 08-1
Chapter Tests 01-11, 02-6, 03-26, 04-10, 05-4, 06-14, 07-14, 08-6
Charging System
    Using 15 Once (0.4L) Containers 08-4
    Using Charging Station 08-3
Checking and Adding Oil
    Axial Piston Compressors 07-7
    Reciprocating Piston Compressors 07-6
Checking Clutch Coil for Electrical 07-12
Checking Evaporator Output 05-3
Checking Relative Temperature at High and
  Low Sides of System 05-3
Checking System for Full Charge 05-3
Choice of Compressor 03-6
Clutch
    Magnetic 03-20
Compound Gauge (Low Side) 04-3
Compressor 03-2
    Choice 03-6
    Noise Complaints 03-7
    Relife Valve 03-4
Compressor Noise Complaints 03-7
Condenser 03-7
    Types 03-8
Control
    Thermostat 03-19

D

Definitions of Terms and Symbols A-1
Dehydrator 03-16
Diagnosis 03-25
Diagnostic Chart 06-14
Disposal, Recycling Guidelines 02-5

E

Evacuating System
    Charging Station 08-2
Evaporator 03-13
Expansion Valve 03-9
Externally — Equalized Expansion Valve 03-11

F

Fan Speed 03-14
Flooded or Starved Evaporator Coils 03-15
Flow Charts for Diagnosing the System 06-6

G

Gauge and Manifold Set 04-3
Gauge Manifold 04-4
Guidelines for Recycling Disposal 02-5

H

High Pressure Gauge (High Side) 04-4
High- and Low-Pressure Switches 03-6
High-Pressure Switch 03-6
Hot Gas Bypass 03-22

I

Information Sources 02-6
Installing Gauge Set to Check System 07-1
Internally — Equalized Expansion Valve 03-10
Isolating Compressor from System 08-5

L

Leak Detectors 04-9
Leak Testing System Using Electronic 07-13
Lines and Connections 03-25
Location 03-25
Low-Pressure Switch 03-6

M

Magnetic Clutch 03-20
Magnetic Clutch Systems 03-18
Measurement Conversion Chart A-4
Modulator Valve 03-24
Moisture in the System 02-4

O

Operating Inspection of System 05-3
Operation 03-25
Other Service Tools 04-10
Other Terms for Refrigerants 02-1
Other Types of Suction Throttling Regulators 03-24

P

Pressure and Temperature Relationship 02-1
Problems
    Flooded or Starved Evaporator Coils 03-15

Continued on next page

Index

	Page
Problems of Flooded or Starved Evaporator Coils	03-15

R

Receiver-Dryer	03-16
Recycling Disposal Guidelines	02-5
Refrigerant	03-17
Refrigerant Recovery	08-1
Refrigerant Recovery and Recycling Station	04-2
Refrigerants	
Handling	02-2
Other Terms	02-1
Safety Rules	02-3
Regulators	
Suction Throttling	03-23
Relife Valve	
Compressor	03-4

S

Screens in the System	03-17
Sealed Sensing Bulb Expansion Valve	03-9
Service	03-25
Service Precautions	03-12
Service Valves	04-6
Shaft Seal Leak Test	07-5
Shutoff Switch	
Superheat	03-5
Solenoid Bypass	03-22
Speed	
Fan	03-14
Suction Throttling Regulators	03-23
Superheat Shutoff Switch	03-5
System	
Bypass	03-22
Operating Inspection	05-3
Visual Inspection	05-2

T

	Page
Temperature-Pressure Relation Chart -- Low Side	
R-12	01-9
R-134a	01-10
Test Hoses	04-5
Thermostat	03-18
Thermostat and Magnetic Clutch Systems	03-18
Thermostat Control	03-19
Thermostatic Temperature Control Switch	07-11
Troubleshooting Customer Complaints	06-2
Types of	
Suction Throttling Regulator	03-24
Types of Condensers	03-8

U

Use of Screens in the System	03-17

V

Vacuum Pump	04-10
Valve	
Expansion	03-9
Valve, Relief	
Compressor	03-4
Vavle	
Modulator	03-24
Visual Inspection of the System	05-2
Volumetric Test of Compressor	07-4

W

What Happens When Refrigerant Is Blocked	03-17